Dark Side of the Universe

Dark Matter, Dark Energy, and the Fate of the Cosmos

Dark Side of the Universe

Dark Matter, Dark Energy, and the Fate of the Cosmos

By Iain Nicolson

Canopus Publishing Limited

First published by Canopus Publishing Limited 2007

A catalogue record for this book is available from the British Library

ISBN 0 95498463 3

Artworks	James Symonds
Project Editor	Julian Brigstocke
Copy Editor	Helen Acosta
Proofreader	Sarah Tremlett
Indexer	Bill Johncocks
Editorial Director	Robin Rees

Produced by Canopus Publishing Limited
27 Queen Square, Bristol, BS1 4ND, UK
www.canopusbooks.com

Printed in China

Acknowledgments

I am grateful to all of the many institutions and individuals who have kindly given permission to reproduce images. I would like particularly to thank the following for welcome advice and/or for help in seeking out, supplying and in some cases, specially modifying, images: Pierre Astier, Paul Butterworth, Jean-Charles Cuillandre, Duncan Forbes, Britt Griswold, Gary Hinshaw, Dan Lewis, Julia Maddock, Javier Méndez, Michael Merrifield, Ben Moore, Sean Paling, Saul Perlmutter, Tom Shanks, Nik Szymanek, and Sue Tritton.

On the production side, I would like to thank all of the team at Canopus Publishing Limited – Editorial Director Robin Rees for embracing the project and for his enthusiastic and energetic support throughout, Helen Acosta for her meticulous editing of the text, James Symonds for so skilfully translating my rough sketches and ideas into high quality artworks and diagrams, Henry Rees for additional assistance with the acquisition of images and Project Editor Julian Brigstocke for his Herculean efforts in putting it all together.

Above all, it's the greatest of pleasures to pay tribute to all of the army of astronomers, cosmologists and physicists – observers, experimentalists and theoreticians alike – whose work has revolutionised our view of the cosmos and who are striving determinedly to unveil the physical nature of the dark side of the universe.

Iain Nicolson
October 2006

Contents

A zoom lens in space
This massive cluster of galaxies, called Abell 1689, contains a trillion stars and a huge amount of dark matter. Its gravity acts as a giant lens that bends and magnifies the light of galaxies far beyond it.

CHAPTER ONE

Anatomy of the Cosmos

The scale of the universe, in space and time, seems overwhelmingly great compared to the scale of everyday objects here on Earth, or to the brief span of a human life. Yet matter also exhibits a rich variety of structure on scales which are almost inconceivably tiny by comparison with everyday objects. Whereas astronomers look to ever-greater distances to map the distribution of matter in the universe, particle physicists probe the tiniest of scales in their attempts to unravel the microscopic structure of matter, and to understand the forces that control its behaviour. Cosmologists, who strive to comprehend the structure, origin, evolution and ultimate fate of the universe as a whole, have their feet firmly planted in both camps – for the behaviour of the universe on the largest of scales is intimately related to particles and forces on the tiniest of scales.

Here, in a nutshell, is a brief outline of the content and scale of the universe and the forces that govern its behaviour.

Our cosmic locality - from the Earth to the stars

The Sun is the dominant body in the Earth's neck of the woods. A typical star, the Sun is a self-luminous globe of gas – composed mainly of hydrogen and helium – which is powered by nuclear reactions that take place deep down in its core. Its diameter of 1,390,000 kilometres is more than a hundred times greater than that of the Earth, and its volume is sufficiently great to contain well over a million bodies the size of our world. The Earth is a member of the family of planets that revolve around the Sun and which, together with a host of smaller bodies, and quantities of gas and dust, comprise the Solar System.

Our nearest neighbour in space is the Moon – a rocky, airless world, with just over one quarter of the Earth's diameter – which revolves around our planet at an average distance of 384,400 kilometres. This distance is not too hard to visualise, for it is approximately equal to travelling ten times round the Earth's equator. The average distance between the Earth and the Sun – 149,600,000 kilometres – is about four hundred times greater than the distance of the Moon. In order of distance from the Sun, the planets are: Mercury, Venus, the Earth, Mars (these four are rocky worlds), Jupiter, Saturn (both are giant planets, about ten times the size of the Earth, which are composed mainly of hydrogen and helium), Uranus and Neptune (each of which is about four times the diameter of the Earth and is composed of a mixture of gas and slushy ices). Neptune trundles slowly around the Sun at an average distance of 4,495 million kilometres – about thirty times the Earth's distance. More distant still are several dwarf planets, best known of which is Pluto (a world of rock and ice that is smaller than the Moon), and a multitude of tiny icy bodies.

The realm of the stars

The stars lie at distances which are enormously greater than those which separate the planets from the Sun. Our nearest stellar neighbour (apart from the Sun itself) is Proxima Centauri, a dim red star in the southern hemisphere constellation of Centaurus (the Centaur), which is more than a quarter of a million times further away than the Sun. The following scale model may give a feel for the distances involved. If we were to represent the Sun by a small melon, 14 centimetres in diameter, the Earth would be

Galaxies galore

Looking deep into space and far back in time, this Hubble Space Telescope image (called the Hubble Ultra Deep Field) shows a myriad of galaxies. The smallest red dots are some of the most distant galaxies known, so far away that we are seeing them as they used to be when the universe was just 800 million years old.

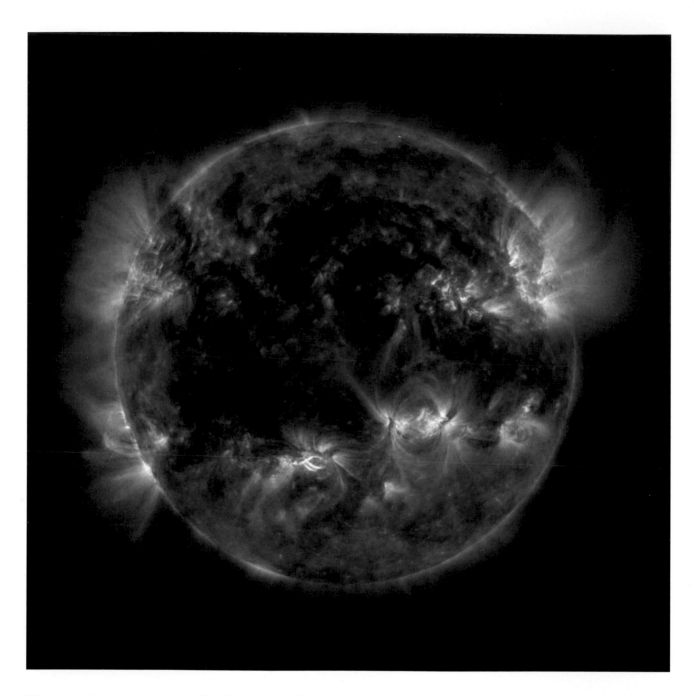

This extreme ultraviolet image of the Sun - our neighbourhood star - shows several bright active regions (where strong magnetic fields are concentrated) together with numerous loop-shaped structures that extend high into the Sun's hot, but exceedingly tenuous, atmosphere.

a small pinhead (1.25 millimetres in diameter) at a distance of 15 metres, and Neptune a small ball-bearing located 450 metres away. On this scale, Proxima Centauri (which is smaller than the Sun) would be an orange, 4000 kilometres away. If the model Sun were in New York, the nearest 'star' would be in San Francisco; if it were in London, the nearest 'star' would be in Egypt.

A useful way of describing distances in the universe at large is to think of how long it

would take for a ray of light to traverse these immense spaces. Light is the fastest-moving entity in the universe, and travels through empty space at a speed of 300,000 kilometres per second. At this speed, a ray of light would take 1.3 seconds to reach the Earth from the Moon, 8.3 minutes from the Sun and about 5.5 hours from Pluto. Proxima Centauri is so far away that its light takes over 4.2 years to reach us. The distance travelled by light in one year (9.46 trillion kilometres, which can be written as 9.46×10^{12} km – see 'Powers of ten', below) provides a convenient unit for distance measurement, which is called the light-year. The distance of Proxima Centauri is 4.2 light-years.

Within the Solar System, distances can be measured with great precision by means of radar: if astronomers transmit a pulse of radio waves – which travels at the speed of light – towards a target planet, and measure the time between the transmission of the signal and the arrival of the returning 'echo', they can work out how far away it is. However, we cannot use this technique with stars. Instead, the fundamental method of measuring their distances relies on parallax – the apparent shift in the position of an object when it is viewed from two different locations. If the position of a nearby star is measured when the Earth is on one side of the Sun and again, six months later, when the Earth has moved to the opposite side of the Sun, the apparent shift in its position will depend on its distance – the greater the distance, the smaller the shift. Half of the maximum overall shift in position is called the annual parallax. Knowing the distance between the Earth and the Sun and the annual parallax

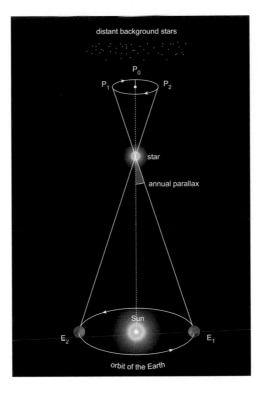

As the Earth moves round its orbit from E_1 to E_2, the apparent position of a star shifts from P_1 to P_2. The annual parallax of the star is its maximum angular shift from its mean position on the sky, P_0.

Powers of ten

A useful shorthand way of writing very large, or very small, numbers is to use index notation, or 'powers of ten', where 10^n ('ten to the power n') represents the number 1 followed by n zeros, and 10^{-n} is the number 1 divided by (1 followed by n zeros). n is the index, or the 'power of 10'. For example, one hundred (100) is 10×10, which is 10^2 (ten to the power two), one thousand (1000) is $10 \times 10 \times 10$, which is 10^3 (ten to the power three), and so on. Conversely, one-tenth (0.1) is 1/10, which is 10^{-1} (ten to the power minus one), one hundredth (0.01) is 1/100, which is 10^{-2} (ten to the power minus two), and so on. More examples are given in the table below:

Number	Decimal form	Index form
one hundred	100	10^2
one thousand	1,000	10^3
one million	1,000,000	10^6
one billion	1,000,000,000	10^9
one trillion	1,000,000,000,000	10^{12}
one hundredth	1/100 = 0.01	10^{-2}
one thousandth	1/1,000 = 0.001	10^{-3}
one millionth	1/1,000,000 = 0.000001	10^{-6}
one billionth	1/1,000,000,000 = 0.000000001	10^{-9}
one trillionth	1/1,000,000,000,000 = 0.000000000001	10^{-12}

A number such as 250 would be written as 2.50×10^2 (2.5 times ten to the power two), and a large number such as 9,460,000,000,000 would be represented by 9.46×10^{12}. In similar fashion, a number such as 2.5 millionths (0.0000025), would be written as 2.5×10^{-6}.

and the value of parallax becomes smaller. Beyond a few hundred light-years, the errors become very substantial and astronomers then have to draw on other, less direct, techniques to measure distances. Nevertheless, parallax provides the first rung on the cosmic 'distance ladder', and is the foundation upon which all other measurements of the scale of the universe ultimately depend.

Stars and their life cycles

Stars exhibit a huge spread of properties. Some are more than a million times as luminous as our Sun (such stars are very rare), while at the other extreme, some have less than one hundred thousandth of the Sun's luminosity. Whereas the Sun has a surface temperature of just under 6000 K, the hottest stars have temperatures in excess of 100,000 K, and the coolest, in the region of 1000 K. (The 'Kelvin', or 'Absolute' temperature scale begins at Absolute Zero – the lowest possible temperature – which corresponds to –273°C. The unit of temperature, which is equivalent to one degree on the Celsius scale, is called the kelvin, and is denoted by the symbol 'K'. On this scale, 0 K corresponds to –273°C, 273 K to 0° C, 373 K to 100° C, and so on.) The largest stars are hundreds or even thousands of times bigger than our Sun, but at the other end of the scale there are highly compressed stars called white dwarfs, which are similar in size to the planet Earth. Even more extreme, though rare, are neutron stars, which are little larger than a major city, yet contain as much material as the Sun.

Between the stars lies an exceedingly tenuous, clumpy mixture of gas (mainly hydrogen) and tiny particles of dust. If a cloud contains one or more extremely hot and highly-lumi-

Orion (the Hunter) is one of the most imposing constellations in the sky. The three stars close to the centre of this image make up Orion's 'belt'. The small fuzzy patch below the belt is the Orion Nebula, the bright star high above the left-hand end of the belt is the red supergiant, Betelgeuse, and the bright star below the right-hand end of the belt, near the bottom of the image, is the blue-white supergiant, Rigel.

of the star, astronomers can use simple trigonometry to calculate the distance of the star.

Stellar parallaxes are always very small angles. For example, the annual parallax of Proxima Centauri – the nearest star – is just 0.772 seconds of arc (one second of arc – abbreviation arcsec – is one sixtieth of one sixtieth of a degree). Measuring such tiny angles is a difficult task, and the observational errors in parallax measurements become more and more significant as the distance increases

The parsec

Because there is a straightforward link between the distance of a star and its parallax – the smaller the parallax angle, the more distant the star – astronomers have devised another unit of distance measurement that takes this relationship into account. The parsec is a unit of distance equal to the distance at which a star would have an annual parallax of precisely one second of arc. It is equivalent to 3.26 light-years, or 206,265 times the average distance between the Earth and the Sun. Proxima Centauri, with an annual parallax of 0.772 arcsec, lies at a distance of 1.3 parsecs (1/parallax = 1/0.772 = 1.295), which is equivalent to 4.2 light-years or about 270,000 times the distance between the Earth and the Sun.

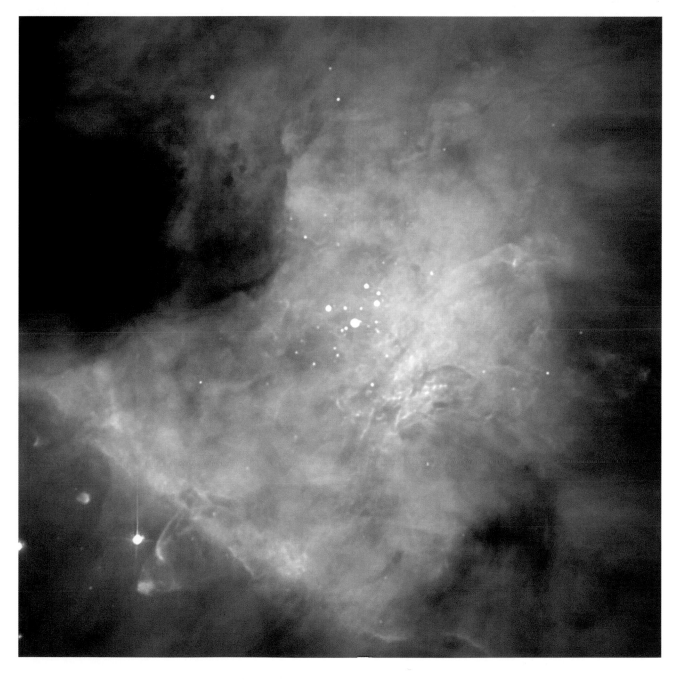

nous stars, the intense ultraviolet radiation emitted by those stars will cause the surrounding gas to glow, giving rise to a misty patch of light that is called an emission nebula (nebula – plural nebulae – is the Latin word for cloud). Other clouds emit radio waves and may therefore be detected even if they do not emit visible light. Massive, cool, dust-laden clouds are where new stars are born.

A star begins to form when a clump of gas which is denser than its surroundings starts to fall together under the influence of gravity. As the cloud collapses, its internal temperature rises. Eventually, when the temperature has risen to about 10 million kelvins, atomic nuclei collide so violently that fusion reactions begin to take place – reactions that convert the lightest chemical element (hydrogen) into the next-lightest (helium) and which liberate copious quantities of energy. Once a newly formed star has reached this stage, the outward pressure exerted by the hot gas in its interior balances

The Orion nebula is a glowing gas cloud composed mainly of hydrogen. A typical emission nebula, it is caused to shine by ultraviolet light emitted by the group of four hot young stars, known as 'the Trapezium', which lie at its centre.

the inward pull of gravity and the star ceases to contract. It becomes a stable 'main sequence' star, which shines thereafter, with very little change, for most of its life.

The key to the process that powers the stars is a relationship between mass and energy that emerged from Albert Einstein's special theory of relativity, which was published in 1905 and which revolutionised our ideas about space, time, motion and energy. Einstein showed that energy (E) and mass (m) are equivalent, and are related to each other, and to the speed of light (c), by that most famous of equations, $E = mc^2$ (energy equals mass multiplied by the square of the speed of light). If a quantity of matter is 'destroyed', the amount of energy that is released is given by Einstein's equation. Because the speed of light is a large number, and the speed of light squared a very large number, a colossal amount of energy is released when even a modest amount of matter is 'destroyed'. For example, if one kilogram of any kind of matter (a lump of coal, a bag of sugar, or a copy of this book) were to be converted completely into energy, the amount of energy that would be released would be enough to power a 1-kilowatt electric heater for nearly 3 million years.

In one second, a star like the Sun converts some 600 million tonnes of hydrogen into helium. During the process a small fraction of that mass – just over 4 million tonnes – is converted into energy. Despite this prodigious

consumption of 'fuel', astronomers reckon the Sun has been shining in this way for nearly 5 billion years, and has sufficient reserves of hydrogen to keep it going for at least another 5 billion years.

Like everything else, however, a star has a finite lifetime. When the stock of hydrogen in its core eventually runs out, the core is squeezed inwards by the weight of the star's outer layers. The outer regions of the star swell up causing the star to expand to tens or even hundreds of times its previous size, becoming a so-called red giant (see 'The Hertzsprung–Russell diagram', right). At this stage a new nuclear reaction – which converts helium to carbon and oxygen – switches on in its shrunken core. The star rapidly consumes its remaining reserves of fuel, then sloughs off its outer shell, and shrinks down to become an extremely dense and compact object – a white dwarf – which eventually, over many aeons of time, cools down and fades to obscurity. This is the fate which awaits the majority of stars, the Sun included.

High-mass stars (those relatively rare stars which have ten to a hundred times as much mass as the Sun) end their lives in a much more spectacular fashion. In the core of a high-mass star, nuclear reactions produce a succession of heavier elements up to and including iron. Once the core has turned into iron it cannot generate any more energy by means of fusion reactions, and its fate is sealed. Unable to support itself against gravity, the core abruptly collapses, releasing so much energy that most of the star's material is blasted into space in a catastrophic explosion that blows the star apart. The collapsing core, meantime, is crushed to incredible density by the overwhelming force of gravity and becomes an exceedingly compressed body – 10 or 20 kilometres in diameter – which is called a neutron star. In more extreme cases, nothing can resist the crushing force of gravity, and the collapse continues until all of the star's matter is compressed into a point of infinite density, which is called a singularity. Before that stage is reached, the force of gravity at the surface of the collapsing star becomes so great that

This Hubble Space Telescope image shows Sirius, the brightest star in the sky, along with its faint, tiny white dwarf companion, Sirius B (tiny dot at lower left). The cross-shaped spikes and concentric rings round Sirius, and the small ring round Sirius B are instrumental artefacts. Sirius B orbits around Sirius in a period of 50 years.

nothing – not even light – can escape its clutches, and the star disappears from view. The collapsed star becomes a black hole – a singularity surrounded by a region of space where gravity is so powerful that nothing can escape from within it.

A star which blows itself apart in this way, and which flares up briefly to be as bright as hundreds of millions of Suns, is called a Type II supernova. Even more brilliant is another class of supernova, known as Type Ia, which is produced by a very different mechanism (see Chapter 9). As we shall see, supernovae have played, and continue to play, a pivotal role in enabling astronomers to probe the furthest reaches of the universe and thereby to determine how it is evolving with time.

Islands in space

The Sun is a member of a vast island of stars that is called a galaxy. Our Galaxy is a disc-shaped system, about 100,000 light-years in diameter, which contains more than 100 billion stars. Extending outwards from a central bulge where stars are most densely concentrated is a pattern of spiral-shaped 'arms' of material, dominated by hot young stars, regions of star formation and clouds of gas and dust. The Solar System lies close to the plane of the disc

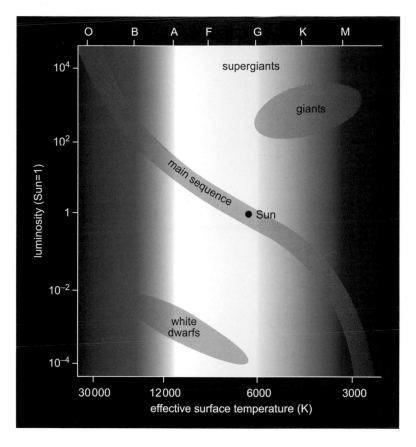

(the galactic plane), some 27,000 light-years from the galactic centre. Viewed from this location, the combined light of millions upon millions of stars in the galactic disc produces

The Hertzsprung-Russell diagram plots the luminosities of stars against their surface temperatures or spectral classes. Most stars lie in the band labelled 'main sequence'. Giants and supergiants lie above the main sequence and white dwarfs below.

The Hertzsprung-Russell diagram

A very useful way of comparing the properties of different kinds of stars is to plot them on a Hertzsprung-Russell (or H-R) diagram. Diagrams of this kind, which were devised in the early part of the twentieth century by the Danish astronomer Ejnar Hertzsprung and the American astronomer Henry Norris Russell, plot luminosity (light output) on the vertical axis and surface temperature (or an equivalent quantity such as colour, or a classification based on stellar spectra) on the horizontal axis. Luminosity increases from bottom to top along the vertical axis and temperature decreases from left to right along the horizontal axis. When large numbers of stars are plotted, according to their luminosities and temperatures, on an H-R diagram, the majority lies within a band, called the main sequence, which slopes from upper left (high luminosity, high temperature) to lower right (low temperature, low-luminosity). The Sun is a typical main sequence star.

Some stars lie above and to the right of the main sequence. These stars, which are larger and more luminous than main sequence stars with the same surface temperature, are called giants (or in extreme cases, supergiants); prominent among these stars are the red giants, relatively cool stars which are hundreds or thousands of times as luminous as the Sun. Below and to the left of the main sequence is a batch of stars with high surface temperatures but low luminosities, which are known as white dwarfs. During its lifetime, a typical star will firstly become a main sequence star (spending most of its life on the main sequence), and will then expand to become a red giant for a time, finally shrinking down to become a white dwarf.

(Left) NGC 4565 is an edge-on spiral galaxy which lies at a distance of some 30 million light-years. Its yellowish central bulge juts out above and below the lanes of dark dust that permeate its flattened disc. Viewed edge-on, our own galaxy (the Milky Way Galaxy) would look rather like this.

(Right) Spiral galaxy M83 lies at a distance of about 15 million light-years. The image shows spiral arms that are rich in hot young stars, and a complex pattern of dark dust lanes. It gives a good impression of what our Galaxy would look like if viewed, face-on, from afar.

a faint band of misty light, called the Milky Way, that extends across the sky from horizon to horizon, and which can be seen with the unaided eye on a clear, dark, moonless night. For this reason, astronomers call the galaxy to which the Sun belongs, the Milky Way Galaxy (for historical reasons it is also sometimes known as 'the Galaxy', with a capital 'G').

Far beyond the confines of the Milky Way Galaxy, lie billions of other galaxies. Some are spiral, like our own, some are elliptical (oval), and others are irregular, having no well-defined shape or structure. The nearest large spiral galaxy is located in the constellation of Andromeda. Otherwise known by its catalogue number of M31 (object number 31 in the catalogue of fuzzy, 'nebulous' objects published in 1781 by the French astronomer, Charles Messier), the Andromeda galaxy is similar to, but larger than, our own Galaxy, and lies at a distance of 2.5 million light-years.

Despite its great distance, under ideal conditions the central part of the Andromeda galaxy can be glimpsed by the unaided eye as a faint misty patch of light. It is by far the most distant object that can be seen by the naked eye. When looking at the Andromeda galaxy we are gazing back in time – seeing it as it was, 2.5 million years ago, when the light we are now receiving set off on its journey across the intervening void of space.

In addition to 'normal' galaxies such as our own, there are various species of 'active galaxy' which emit far more energy – across a wide range of wavelengths – than ordinary galaxies. Most of their energy pours out of an intensely luminous compact central nucleus – called an active galactic nucleus, or AGN – from which emerge, in many cases, jets of fast-moving particles. Each AGN is believed to be powered by energy which is released when matter spirals in towards a supermassive black hole – weighing from tens of millions to several billion times the mass of the Sun – that lies at its centre. Among the most luminous active galaxies are objects called quasars. The name derives from 'quasi-stellar radio source' – a reference to the fact that the first objects of this kind to be identified were strong emitters of radio waves and had central nuclei that were almost star-like in appearance. Because of their extreme brilliance, quasars can be seen at immense distances.

Galaxies are gregarious

Galaxies are distributed in a clumpy fashion. Some, like our own, are members of small groups that contain a few, or a few tens of members. Others are contained within clusters that contain hundreds or thousands of galaxies, spread over volumes of space around 10 million light-years in diameter. Located at a distance of about 50 million light-years, the Virgo cluster, which lies in the direction of the constellation Virgo and contains about

Abell 1185 is a large galaxy cluster located at a distance of about 400 million light-years, which provides a showcase for the main types of galaxies: ellipticals, spirals and irregulars. In densely populated clusters like these, close encounters between galaxies are frequent events. At the bottom of this image, such an event is caught in progress. Stars, gas and dust are being torn from each galaxy by mutual gravitational forces. Image courtesy of Canada-France-Hawaii Telescope/J-C Cuillandre/Coelum.

The central portion of the Virgo cluster, some 50 million light-years distant, contains many elliptical and spiral galaxies.

This image shows part of the 'Hubble Deep Field North', a region of sky about one thirtieth of the Moon's apparent diameter. Some of the remote galaxies, which appear as tiny dots, are so far away that their light has taken more than ten billion years to reach us.

a thousand member galaxies, is the nearest major cluster. On an even larger scale, clusters and groups of galaxies are aggregated together into loose, sprawling structures called superclusters, which typically have diameters in the region of 100–200 million light-years. Our Local Group of galaxies lies on the outer fringe of the Virgo supercluster, which is centred on the Virgo cluster. Although it spans 100 million light-years and contains some 5000 galaxies, the Virgo supercluster is nevertheless a relatively modest one.

Overall, the distribution of luminous matter in the universe is rather 'frothy'. Galaxies are gathered together into clusters, superclusters, long straggly filaments and sheets, or 'walls', which can be as much as 500 million light-years across, yet only tens of millions of light-years thick. These huge aggregations of matter are

separated by, and wrap themselves around, great voids, 100–200 million light-years across, within which luminous galaxies are almost completely absent. Precisely why the overall distribution of matter should resemble an aggregation of soap bubbles, or a Gruyère cheese if you prefer a more edible analogy, is one of the key problems in modern cosmology. As we shall see in later chapters, luminous matter is only a small part of the story.

One consequence of the finite speed of light is that we see remote objects not as they are 'now', but as they were at the instant when the light that we are presently receiving departed from those objects to begin its journey across the intervening space. Thus, we see the Moon as it was 1.3 seconds ago, the Sun as it was 8.3 minutes ago, Proxima Centauri as it was 4.2 years ago, the Andromeda galaxy as it was 2.5 million years ago, and so on. When we look at distant galaxies, millions or billions of light-years away, the 'look-back' time (how far back in time we are looking when we observe these remote objects) is such that we are seeing them as they were millions or billions of years ago. The most remote and luminous quasars are so far away that we are seeing them as they were some 12–13 billion years ago.

Measuring the outer limits

Establishing the distance scale of the universe has been one of the great triumphs of observational astronomy – a task fraught with difficulty, as astronomers have sought to improve the precision and range of their measurements in the face of known, and sometimes unknown, errors of observation.

Fundamentally, the two key methods for

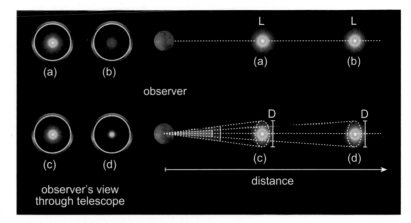

measuring large distances are 'luminosity distance' and 'diameter distance'. The first relies on the fact that the apparent brightness of a source of light decreases in proportion to the square of its distance. If the distance is doubled, the apparent brightness reduces to one quarter of its previous value (assuming no light is absorbed or scattered away en route to the Earth). If we know the inherent luminosity of an object, such as a particular type of star, we can work out its distance – and hence the distance of the galaxy within which it is embedded – by comparing its apparent brightness in the sky (the amount of light arriving from the star) with its true intrinsic luminosity (the amount of light it is emitting from its surface).

Diameter distance relies on knowing the true physical diameter of a distant object and comparing this to its observed apparent (angular) diameter. Try holding a coin at different distances. Its apparent size decreases as its distance increases; if the distance is doubled, the apparent size of the coin reduces to half its previous value. If we know the physical size of an object, we can work out how far away it must be in order to look as small as it does. If astronomers have a good idea of how big a particular class of object is (for example, a cluster of stars, a luminous nebula, or the jet of gas emanating from the nucleus of an active galaxy) they can estimate the object's distance by comparing its apparent angular size to its assumed physical size.

Best known of the astronomers' inventory of 'standard candles' (objects of known luminosity) are Cepheid variables, stars which are thousands of times more luminous than the Sun and which can readily be identified because they vary in brightness in a regular, periodic way. As American astronomer Henrietta Leavitt first demonstrated in 1912, from a study of Cepheid variables in the Small Magellanic Cloud (a satellite galaxy of the Milky Way which is visible to the naked eye in the southern hemisphere sky), there is a well-defined relationship – known as the period-luminosity law – between the luminosity of a Cepheid and the period of time over which it varies in brightness. The more luminous the Cepheid, the longer its period.

Despite their considerable luminosities, Cepheids cannot be detected at distances much greater than that of the Virgo cluster (50–100 million light-years), and astronomers need to use different kinds of standard candles to probe the remoter recesses of the universe. Pre-eminent among the current crop of standard candles are supernovae (exploding stars). The downside to using supernovae is that they are rare events – only one or two occur per century in a typical galaxy.

Investigating the nature of stars and galaxies

Astronomers study the universe by detecting and analysing light, particles and other kinds of radiation that arrive at the Earth from distant objects.

Light is a form of electromagnetic radiation – an electric and magnetic disturbance that travels through space at a speed of 300,000 kilometres per second (the velocity of light in a vacuum). We can think of light as a wave motion, analogous to a wave on water. The

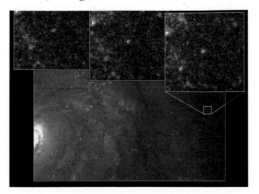

Luminosity distance (upper): if two stars are equally luminous but one (b) appears only a quarter as bright as the other (a), then (b) must be twice as far away as (a). Diameter distance (lower): if two objects have the same physical diameter (D) the one that looks smaller (d) must be further away than the one with the larger angular diameter (c).

This Hubble Space Telescope image (lower) of part of the spiral galaxy M100 (a member of the Virgo cluster) shows the location of a Cepheid variable (identified by the small inset box) in one of the galaxy's spiral arms. The three frames at the top show changes in the brightness of this variable star, which is located at the centre of each frame, over a period of three weeks.

Supernova 1987A flared up in the Large Magellanic Cloud (a satellite galaxy of the Milky Way which lies at a distance of 170,000 light-years) in February 1987. Despite its great distance, it became bright enough to be visible to the unaided eye. The two images show the appearance of part of the Large Magellanic Cloud before (upper), and (lower) at the time when the supernova was seen. The supernova is the exceptionally bright star near the bottom right-hand corner of the lower image.

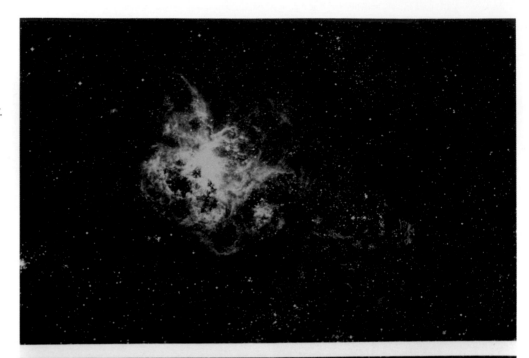

distance between two successive wavecrests is the wavelength, and the number of wavecrests that pass a particular observer in one second is the frequency. Because all light waves travel at the same speed (in a vacuum), more wavecrests of short-wave light will pass by in one second than wavecrests of long-wave light. The shorter the wavelength, the higher the frequency.

Visible light spans a range of wavelengths from just under 400 nanometres (a nanometre, symbol nm, is a billionth of a metre) to around 700 nm, and our eyes respond to different wavelengths by perceiving different colours. For example, blue light has a wavelength of around 400 nm, yellow light about 550 nm, and red light about 700 nm. Wavelengths shorter than visible are known as ultraviolet, whereas

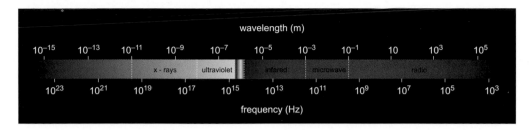

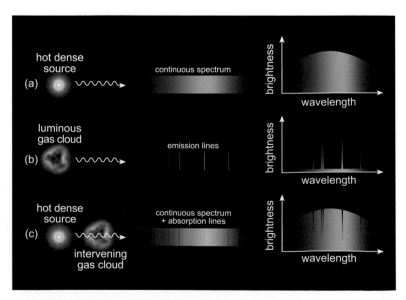

The electromagnetic spectrum extends from the shortest wavelength gamma-rays to the longest radio waves. Visible light (the central rainbow band of colours) makes up only a tiny portion of the full range of electromagnetic radiation.

(a) A hot dense source of light emits a continuous spectrum which, in the visible region of the spectrum, appears as a rainbow band of colours. (b) A hot cloud of low-pressure gas radiates at certain wavelengths only, to give an emission line spectrum. (c) If a continuous spectrum passes through a low-pressure gas cloud, dark absorption lines are superimposed on the spectrum. The graphs on the right show how the brightness of each type of spectrum varies with wavelength.

those which are longer than visible are called infrared. The complete range of electromagnetic waves (which is called the electromagnetic spectrum) is divided into a number of bands which, from the shortest to the longest wavelengths are labelled: gamma-ray, x-ray, ultraviolet, visible, infrared, microwave and radio. Radio waves have wavelengths of metres or more, whereas gamma-rays have wavelengths of less than 0.01 nanometres.

When a beam of white light (a mixture of wavelengths) passes through a glass prism (or an equivalent device called a diffraction grating), the different wavelengths are refracted (deflected) by different amounts, and the beam is spread out into a rainbow band of colours from violet to red, which is called a continuous spectrum. The spectrum of a typical star consists of a continuous spectrum on which are superimposed numerous dark lines. When dark lines in the spectrum of the Sun were discovered in the early part of the nineteenth century, no-one knew how they were caused or what they signified. In 1859, however, Gustav Kirchhoff and Robert Bunsen showed that a hot, dense body (solid, liquid or gaseous) emits a continuous spectrum, whereas a hot cloud of low-density gas emits light at certain wavelengths only, with the result that its spectrum consists of a number of bright lines (emission lines), each line having its own particular wavelength. Bunsen and Kirchhoff also demonstrated that if a continuous spectrum passes through a cooler, more rarefied cloud of gas, light is absorbed at certain particular wavelengths, thereby imprinting a pattern of dark (absorption) lines on the rainbow band of the continuous spectrum. In similar fashion, absorption lines are imprinted on a star's continuous spectrum as its light propagates out through its tenuous atmosphere.

Each chemical element produces its own characteristic pattern of lines. By identifying the various patterns of lines present in the spectrum of a star, astronomers can determine the chemical composition of stars, gas clouds and galaxies. In practice, analysing and interpreting a spectrum is a complex process, because the detailed appearance of a spectrum depends not only on chemical composition but also on factors such as temperature, pressure, rotation, the presence of magnetic fields, and so on. But, because all these factors exert an influence on a spectrum, astronomers can extract a wealth of information about the chemical and physical properties of stars, gas clouds and galaxies.

Astronomers measure the speed at which a star or galaxy is approaching or receding by utilising a phenomenon known as the Doppler effect – an effect that was first discussed in the context of sound waves by the Austrian scientist Christian Doppler in 1842, and which is very familiar in everyday life. The pitch of a sound depends on the number of

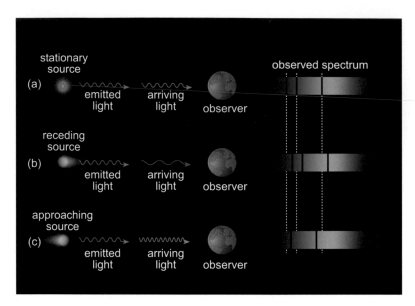

If a source of light is receding, the light that reaches the observer is stretched to a longer wavelength than would be the case if the source were stationary. Lines in the spectrum of that source are displaced (redshifted) towards the longer-wave end of the spectrum. If the source is approaching, spectral lines are displaced (blueshifted) toward the short-wave end of the spectrum. This phenomenon is called the Doppler effect.

waves impinging on your ear every second. If the source of a sound is approaching, more wavecrests per second enter your ear, and the perceived pitch is higher, whereas if the source is receding, fewer wavecrests per second enter your ear and the perceived pitch is lower. A familiar example is the change in pitch of a siren as an emergency vehicle rushes by.

A similar phenomenon occurs with light waves. If a source is approaching, the wavecrests are squeezed closer together, shortening the wavelength and increasing the frequency. Conversely, if the source is receding, wavelengths are stretched and the frequency is reduced. As French astronomer Hippolyte Fizeau showed in 1848, the Doppler effect also changes the wavelengths of spectral lines. If a source of light is stationary (neither approaching nor receding) each line has its own characteristic wavelength, which is called its restwavelength. If a light source is approaching, the wavelength of each line is reduced, and the whole pattern of lines is shifted towards the short wavelength (blue) end of the spectrum; this phenomenon is called a blueshift. Conversely, if the source is receding, the lines are shifted towards the long-wave (red) end of the spectrum, and the phenomenon is called a redshift.

The shift in a line's wavelength is proportional to the speed at which the light source is approaching or receding (the radial velocity)

– the higher the radial velocity, the greater the shift. Consequently, astronomers can measure the speed at which a star, gas cloud or galaxy is approaching or receding by comparing the observed wavelengths of known spectral lines with their rest-wavelengths.

Atoms and spectra

Our present knowledge of the way in which absorption and emission lines are formed, and of why different chemical elements have distinctive sets of spectral lines, stems from a revolution in our understanding of the nature of light, energy and the structure of the atom which began during the first decade of the twentieth century.

In 1900, in order to overcome a serious problem in accounting for the way in which hot bodies radiate energy, which had troubled physicists in the late nineteenth century, German physicist Max Planck proposed that energy is conveyed in discrete little packets, called quanta (singular – quantum). Based on this idea, light can be envisaged as a stream of 'particles', called photons, the energy of a photon being inversely proportional to wavelength – the shorter the wavelength, the higher the energy. For example, compared to photons of visible light, gamma-ray photons are much more energetic, whereas radio photons carry much less energy. It seems that light behaves in some respects like a wave motion, but in others like a stream of tiny particles.

During that same decade, experimenters found that the atom consists of a tiny but heavy, positively-charged nucleus surrounded by a cloud of lightweight negatively-charged particles called electrons; and in 1911, New Zealand-born physicist, Ernest Rutherford, suggested that the atom is like a miniature Solar System, with electrons orbiting around a nucleus rather like planets around the Sun.

Danish physicist Niels Bohr brought these two concepts together in 1913. According to the Bohr model of the atom, electrons can orbit around an atomic nucleus only in certain permitted orbits or shells, each of which corresponds to a particular energy level. If an electron drops down from a higher level (a

Waves versus particles
An example of the wave behaviour of light is the phenomenon of interference, which occurs when two beams of light meet up with each other: if the waves are in phase, so that wavecrests of one wave coincide with wavecrests of the other, the two will add together to produce a 'higher' wave (one of greater amplitude), whereas if the crests of one coincide with the troughs of the other, the waves will cancel out. An example of the particle aspect of light is the photoelectric effect whereby an individual photon, if it has sufficient energy, will cause an electron to be expelled from the light-sensitive material. Electrons will be emitted only if the wavelength of the incoming beam of light is shorter than a certain minimum value (corresponding to the minimum photon energy needed to dislodge the electron). Light of longer wavelength will not dislodge the electrons – no matter how bright the beam – because long-wave photons have insufficient energy. Einstein explained the photoelectric effect in this way in 1905 and was awarded the 1908 Nobel Prize for Physics for this work.

larger orbit) to a lower one (a smaller orbit), it emits a photon of light with energy equal to the energy gap between the two levels. Conversely, if an electron in a lower level absorbs a photon with the right amount of energy, it will jump up to a higher one. Consequently, atoms of a particular element absorb or emit light at certain particular wavelengths only. The wavelength of the absorbed or emitted photon depends on the energy difference between the two levels – the greater the energy gap, the more energetic the photon and the shorter its wavelength. The chemical identity of an atom, and the permitted energy levels of its electron shells, are determined by the mass of, and the total electrical charge attached to, its nucleus. Because the wavelengths at which light is absorbed or emitted depend on the energy differences between these levels, each chemical element produces its own characteristic set of spectral lines.

Looking deep inside the atom

By 1932, physicists had established that atomic nuclei were composed of two different types of particle – protons, which are massive particles with a positive electrical charge (equal in magnitude, but opposite in sign, to the charge on the electron), and neutrons, which are marginally more massive than protons but have zero electrical charge. The number of orbiting electrons in a complete atom, which is electrically neutral, is the same as the number of protons in its nucleus. For example, the nucleus of a hydrogen atom consists of a single proton, and a complete hydrogen atom is com-

posed of a single proton and a single orbiting electron, whereas a helium nucleus contains two protons and two neutrons, and a complete helium atom has two orbiting electrons.

Whereas the diameter of an atom (the overall diameter of its cloud of orbiting electrons) is about 10^{-10} metres (less than a thousandth of the wavelength of visible light), the diameter of its nucleus is about ten thousand times smaller (in the region of 10^{-14} metres). Most of the interior of an atom is 'empty', for the volume of the nucleus – which contains more than 99.9 percent of the atom's mass – is only one trillionth of the total volume of the atom. To put it in context, if an atom were the size of a football ground, its nucleus would be a grape placed on the centre spot.

The electron appears to be a stable and truly fundamental particle (one which is not composed of smaller components), but protons

The Bohr model of the hydrogen atom explains the origin of spectral lines. An electron can jump up from one of the lower permitted energy levels (or 'orbits') to a higher one if it absorbs a photon of the correct energy and wavelength. This process produces dark absorption lines in a spectrum. When electrons drop down to lower energy levels, photons of certain particular energies and wavelengths are emitted. This process produces an emission line spectrum.

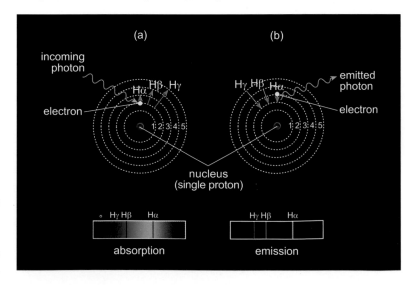

and neutrons are composed of smaller point-like particles called quarks; each consists of a bunch of three quarks – particles whose existence was confirmed experimentally in 1968. The proton is an exceedingly long-lived particle, unlikely, on average, to decay into anything else in less than about 10^{31} years. An isolated neutron, however, will decay – typically within 20 minutes – into a proton, an electron and another kind of particle, called a neutrino (strictly, an antineutrino – see later), which has zero electrical charge and an exceedingly tiny mass.

Over the past half century, physicists have discovered a bewildering array of subatomic and elementary particles, each of which is char-

acterised by its mass, electrical charge and spin (an elementary particle can be visualised as a tiny spinning ball; it is not really like that, but this is a convenient way to visualise the property called 'spin'). Each particle has an equivalent antiparticle, which has the same mass, but opposite charge. For example, the antiparticle of the electron, which is called a positron, has the same mass as an electron but a positive electrical charge. Likewise, an antiproton has the same mass, but opposite (negative) charge to the proton. When a particle and antiparticle are created together as a pair (which happens, for example, when two sufficiently energetic photons collide), the antiparticle will spin in the opposite direction to the particle. Particles and antiparticles have a pathological antipathy towards each other: when a particle collides with its antiparticle, they annihilate each other and convert to a quantity of energy (typically in the form of a pair of energetic photons) equivalent to the combined mass of the particle–antiparticle pair.

The forces of nature

In the present-day universe, the behaviour of matter and radiation is controlled by four fundamental forces: gravity, the electromagnetic force, the weak nuclear force and the strong nuclear force. Gravity controls the behaviour of matter on the large scale – the orbits of planets, the motions of stars, the mutual interactions between galaxies – and plays a key role in the evolution of the universe as a whole. It is a long-range force: according to Newton's law of gravitation the force of gravity diminishes with the square of distance (if the distance is doubled, the force dwindles to a quarter of its previous value) and, although it declines rapidly with increasing distance, the gravitational attraction of a lump of matter does not drop to zero until the distance from it is infinitely great.

The electromagnetic force controls the motion of electrically-charged particles and the emission and absorption of light and other forms of electromagnetic radiation. Whereas gravity is always attractive (two masses always attract each other), the electromagnetic force can be attractive or repulsive. Opposite charges

An atom consists of a nucleus surrounded by a cloud of electrons. The nucleus is composed of protons and neutrons (collectively called 'nucleons') which, in turn, are composed of 'up' and 'down' quarks.

atom — electron — 10^{-10} m

nucleus — proton — neutron — 10^{-14} m

nucleons — quarks

(positive and negative) attract, whereas like charges repel each other. The electromagnetic force enables the positively-charged nuclei of atoms to hold on to negatively-charged orbiting electrons. It too is a long-range force. Like gravity, its strength diminishes with the square of distance.

The strong and the weak nuclear interactions are short-range forces which exert their influence over distances no greater than the diameter of an atomic nucleus. The strong force binds together the nucleons (protons and neutrons) which comprise the nucleus of an atom, and prevents the protons (which all have positive charges and which, therefore, tend to repel each other) in a nucleus from flying apart. At a deeper level, the strong interaction binds together the quarks of which protons and neutrons are composed. The weak nuclear interaction governs the process of radioactive decay whereby unstable nuclei change from one type of nucleus into another, by emitting electrons (beta decay) or more massive clusters of protons and neutrons, called alpha particles.

Particles which are composed of quarks, and which respond to the strong nuclear interaction, are known collectively as hadrons, but are subdivided into baryons – particles, such as protons and neutrons, which are composed of three quarks – and mesons, which are composed of quark–antiquark pairs. Particles which are not influenced by the strong nuclear force (such as photons, electrons or neutrinos) are called leptons. Matter composed of protons and neutrons (the building blocks of atomic nuclei) is known as baryonic matter – it is the stuff of which planets, stars and people are made.

Of the four forces, gravity is by far the weakest (even though it may not seem so if you fall off a cliff!). When measured between two protons, the strong nuclear force is nearly 10^{40} times stronger than their mutual gravitational attraction and about a hundred times stronger than the electromagnetic force (easily strong enough to prevent like-charged protons from flying apart). The weak nuclear force is about a hundred thousand times feebler than the strong, but is still at least 10^{34} times stronger than gravity. However, at distances greater than about 10^{-15} metres, the weak and strong

Mass, energy, charge and spin

Because mass and energy are equivalent to each other, particle physicists tend to express the masses of elementary particles in units of energy. Their preferred unit is the electron volt (symbol eV), which is the energy that an electron acquires when it is accelerated by a potential different (voltage) of 1 volt. It is a tiny unit, equivalent to 1.602×10^{-19} joules (in one second a 100 watt domestic light bulb emits a quantity of energy equivalent to about 600 million million million electron volts); common multiples include keV (a thousand electron volts), MeV (a million electron volts) and GeV (a billion electron volts). Expressed in these terms, the mass of an electron is 0.511 MeV/c^2 (mass = energy divided by the square of the speed of light), which is equivalent to 9.1×10^{-31} kilograms, and the mass of a proton is 938.3 MeV/c^2 (which equates to about 1.67×10^{-27} kilograms).

The unit of charge is equal to the amount of electrical charge carried by the electron. The negatively-charged electron has a charge of -1, whereas the positively-charged proton has a charge of +1. Quarks have fractional electrical charges (2/3 or 1/3 of the charge on the electron) and, forsaking natural economy, come in six different brands, or 'flavours' labelled (in the slightly whimsical fashion that particle physicists are prone to adopt) 'up', 'down', 'strange', 'charm', 'bottom' and 'top'. The up quark has a charge of +2/3, whereas the down quark has a charge of -1/3. A proton consists of two up quarks and one down quark, this combination providing a net charge of +1 (2/3 + 2/3 - 1/3 = +1), whereas a neutron consists of two downs and an up, which gives a net charge of zero (-1/3 - 1/3 + 2/3 = 0).

Spin is expressed in units such that electrons, protons, neutrons and quarks all have spin values of one half (1/2). By contrast, photons have spin -1. Of the veritable zoo of particles, those which have half-integer values of spin (1/2, 3/2 and so on) are called fermions, whereas particles with integral spin values (0, 1, 2 ...) are called bosons. The two different spin families behave in very different ways.

forces have negligible influence. Furthermore, because most bulk quantities of matter in the universe are electrically neutral (they have the same number of positive and negative charges), gravity – despite its inherent weakness – is the force that controls the motions of planets, stars and galaxies.

The bizarre quantum world

While the 'Solar System' model of the atom provides a handy mental picture, it should not be taken too literally. A more complete picture of the nature of atoms is provided by quantum mechanics – one of the great theoretical triumphs of twentieth-century physics – which incorporates several key concepts into our view of nature. First, energy is packaged in tiny units called quanta. Secondly, wave-particle duality: not only do electromagnetic waves behave in some respects like particles (photons), but particles of matter exhibit some wave-like properties, too. Thirdly, Heisenberg's Uncertainty Principle, which implies that the microscopic world of subatomic and elementary particles is a world of probabilities, not certainties.

Propounded by German physicist, Werner Heisenberg, in 1927, the Uncertainty Principle states that we cannot simultaneously know the precise position and momentum (quantity of motion) of a particle. For example, in order to measure the position of a moving electron, we need to illuminate it with light; but each time a photon strikes the electron, it gives it a kick, which changes its motion. The more precisely we try to measure its position, the more we disturb its motion and the less certain we can be about its direction and speed. The Principle also implies that we cannot know the exact amount of energy contained in a microscopic system of particles, or in a tiny volume of space over an arbitrarily short interval of time. Although we can calculate with great precision the statistical average of the orbits of millions of electrons in millions of atoms, we cannot calculate the precise orbit and location of a single electron in a single atom.

The Uncertainty Principle has had a profound impact on our understanding of the nature of particles, forces and the concept of a vacuum. Einstein showed that mass and energy are equivalent ($E = mc^2$), and that one can be transformed into the other. Particles of matter can be transformed into an equivalent amount of energy, and energy can be converted into particles of matter. Because we cannot know the precise amount of energy in each tiny box of space over exceedingly short intervals of time, the Uncertainty Principle opens up the possibility that the energy level could fluctuate sufficiently to create particles, provided that they vanish again before they can be seen or detected by an observer. The process is rather like overdrawing your bank account but paying the overdraft back before anybody notices! Particles which pop into existence and almost immediately vanish again, and which cannot be seen directly, are known as virtual particles, to distinguish them from 'real' ones, which can be detected, seen and measured. What we think of as the empty vacuum is an ethereal world of virtual particles popping briefly into existence then vanishing without visible trace. As we shall see, the concept of virtual particles plays a profound role in understanding the nature of the fundamental forces.

One of the outstanding achievements of twentieth-century physics was the development of quantum theories of the electromagnetic, weak nuclear and strong nuclear forces in which the force acting between real particles is carried, or 'mediated', by virtual force-carrying particles. In the case of the electromagnetic force, the force-carrying particle is the photon. For example, when two negatively-charged electrons approach each other, virtual photons pass between them to convey the electromagnetic influence that causes them to repel each other. The process is analogous to two skaters gliding across the ice on converging paths. If one throws a heavy medicine ball (the 'force-carrying particle') to the other, the act of throwing the ball will cause the thrower to recoil and the act of catching it will push the other away.

The range of a force is determined by the masses of its force-carrying particles – the higher the mass, the shorter the range. This stems from the Uncertainty Principle: the Uncertainty Principle only permits a massive virtual particle to exist for a very short period of time; therefore it cannot travel very far before it has to vanish again. By contrast, a virtual particle of zero mass could travel an infinite distance. Whereas high-energy virtual photons cannot travel very far, low-energy photons can travel much further. The lowest energy which a photon can have is zero, in which case it could travel an infinite distance before having to vanish. Consequently, although the electromagnetic force decreases with increasing distance, it has a huge range. The weak force is conveyed by particles called intermediate vector bosons (W and Z particles) which have masses nearly a hundred times greater than that of a proton, and which, therefore, can only convey the weak force over exceedingly short distances. The strong force acting between nucleons in the nucleus of an atom is conveyed by virtual mesons, which – although less massive than W and Z particles – also have a very restricted range of influence. When acting between the individual quarks inside a proton or a neutron, the strong force is conveyed by force-carrying particles called gluons.

Frustratingly, the most familiar force of all in the everyday world – gravity – is the only one which, so far, has obstinately refused to be cast in the quantum physics mould. For the time being, the best available theory of gravity is the general theory of relativity, which was devised by Einstein some 90 years ago.

Einstein's theory of gravity

Whereas Newton regarded gravity as a force that acted directly across empty space between individual massive bodies, Einstein treated gravity as a phenomenon that arises because space is distorted (curved) in the presence of massive bodies. Within Einstein's theory, the three dimensions of space (length, breadth and height) and the dimension of time (which Newton had regarded as being completely separate from, and independent of, each other) were intimately linked into a four-dimensional entity called space-time. According to the theory, a massive body distorts space-time in

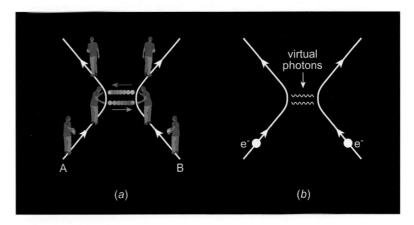

(a)   (b)

its vicinity, and the paths of material bodies and rays of light depend on the curvature of space-time within which they are moving. Thus, a planet travels in its orbit around the Sun, not because it is constrained to do so by a force of attraction acting directly between it and the Sun, but because it is following its natural path in curved space-time. As American physicist Archibald Wheeler famously put it, 'Matter tells space how to curve, and space tells matter how to move'.

The more massive and concentrated the body, the greater the curvature of space-time, and the greater the 'force' of gravity, in its neighbourhood. With increasing distance from the massive body, the curvature becomes less (and the perceived gravitational force weaker). In the absence of any matter (or energy), space-time would be flat (zero curvature), and rays of light would travel in straight lines. A useful analogy is to represent space by an elastic sheet which, in the absence of any massive bodies, is completely flat, like a pool table. A lightweight ball set rolling on a flat elastic sheet will move in a straight line. If a weight were placed on the sheet, it would make an indentation that would deflect the ball from its straight-line path; a heavier weight (representing a greater mass) would cause a larger and deeper indentation, causing greater deflection of the ball. If a ball is set rolling in the right direction at the right speed, it will follow a circular path around the indentation (rather like a 'wall of death' rider), just as a planet pursues its orbit in the curved space that surrounds the Sun.

(a) If two approaching ice skaters throw heavy balls towards each other, the act of throwing and catching causes them to recoil from each other. In an analogous way (b) when two electrons approach each other, the exchange of virtual photons between them conveys the force that causes them to repel each other.

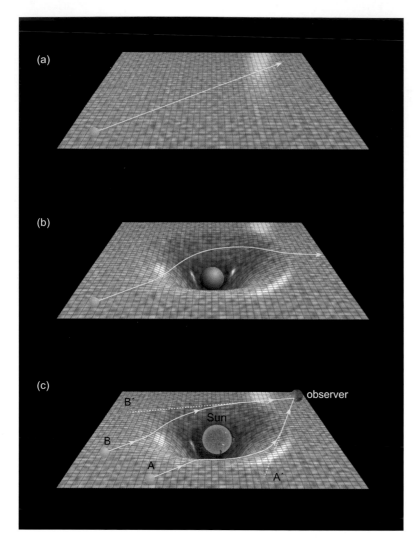

theory of gravity. It also predicted two new phenomena – gravitational redshift (light rays are stretched in wavelength, and photons lose energy, when 'climbing out of' a strong gravitational field), and gravitational time dilation (clocks run slower in strong gravitational fields than in weak ones) – both of which have been confirmed experimentally. So far, general relativity has passed with flying colours every observational and experimental test to which it has been subjected and remains, some 90 years after it was first devised by Einstein, the best available theory of gravity. It is, however, a very different kind of theory from the quantum theories that account for the other three forces of nature.

General relativity remains the pre-eminent theory of space, time and gravity that cosmologists use when attempting to deal with the large-scale properties of the universe as a whole, but quantum physics is the key to handling other forms of force and energy and dealing with the microscopic world of elementary particles. As subsequent chapters will reveal, both approaches play pivotal roles in our attempts to answer fundamental questions relating to the origin, evolution and ultimate fate of the cosmos.

If a photon is represented by a ball rolling across an elastic sheet then *(a)* if the sheet is flat, the 'photon' will move in a straight line; *(b)* a massive body placed on the sheet creates an indentation which deflects the 'photon'. *(c)* The curvature of space in the neighbourhood of the Sun deflects light from stars A and B so that they appear to lie at positions A' and B'.

One of the key predictions of Einstein's theory was that rays of light would be deflected when passing close to massive bodies. This prediction was tested and confirmed during a total eclipse of the Sun which took place on 29 May 1919, by measuring the positions of stars close to the edge of the Sun and comparing them with the positions of those same stars on photographic plates taken when the Sun was not in that region of the sky. In more recent times, the bending of light by gravity has provided astronomers with a valuable tool to measure the distribution of all kinds of matter in the universe, be it luminous or dark (see Chapter 3). General relativity also successfully accounted for an anomaly in the motion of the innermost planet, Mercury, that could not wholly be accounted for by Newton's

Big Bang and Cosmic Destiny

During the early part of the twentieth century, one of the major debating points among astronomers was the question of the nature of the 'extragalactic nebulae' – fuzzy objects, most of which were spiral in shape, which were distributed over the whole sky above and below the plane of the Milky Way. These objects had spectra which resembled the spectra of stars, and were completely different in nature from glowing clouds of gas such as the Orion nebula. On one side were those who contended that the extragalactic nebulae were merely clouds of unresolved stars which lay at the fringes of the giant star system to which the Sun belongs, and who argued that we live in an 'island universe' – a single giant star system ('The Galaxy') surrounded by a sea of emptiness. On the other side were those who reckoned they were independent systems which lay far beyond the confines of the Galaxy.

The issue was resolved in 1923 by Edwin Hubble, who used Cepheid variables (see Chapter 1) to determine the distance of the Andromeda 'nebula'. Although the figure he obtained was only about 40 percent of the presently accepted value (870,000 light-years as opposed to 2,500,000 light-years), it convincingly demonstrated that the Andromeda nebula is a vast independent star system (a galaxy, in modern parlance). Hubble applied the same technique to other extragalactic nebulae and found that they, too, were independent systems. After that, the concept of the single island universe was replaced by the vision of a universe populated with galaxies extending as far as the largest telescopes could see.

A full decade earlier, Vesto Melvin Slipher of the Lowell Observatory, at Flagstaff, Arizona, had begun work on a programme to measure Doppler shifts in the spectra of the extragalactic nebulae. Towards the end of 1912, he succeeded, for the first time, in measuring the Doppler shift in the spectrum of the Andromeda nebula, and found that it was approaching us at a speed of about 300 kilometres per second. By the end of 1923 he had managed to measure Doppler shifts in the spectra of 41 extragalactic nebulae, all but five of which had redshifts which indicated that they were receding from us at speeds of up to

The Andromeda galaxy (M31) is the nearest large galaxy and the most distant object that can be seen with the unaided eye, under ideal conditions. A spiral galaxy, tilted almost edge-on to our line of sight, it lies at a distance of 2.5 million light-years.

The Velocity–Distance
Relation among Extra-
Galactic Nebulae from Edwin
Hubble's original paper
published in Proceedings
of the National Academy of
Sciences on 15 March
1929.

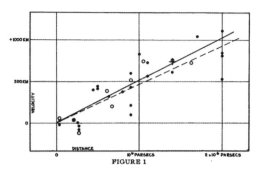

FIGURE 1

1,800 kilometres per second.

Throughout the 1920s, Hubble pressed on with his programme of measuring galaxy distances. Although, with the techniques available at the time, he was unable to detect Cepheids in any but the nearest galaxies, he devised other techniques, such as measuring the brightest luminous objects embedded within galaxies, or the overall brightness and apparent diameters of galaxies as a whole, to push his measurements to progressively greater distances. By combining his own data on galaxy distances with Slipher's Doppler shift data, Hubble found that the measured redshifts in the spectra of the galaxies are proportional to their distances, a result which he set out in an epoch-making paper that was published in 1929. Assuming that the redshifts are due to the Doppler effect, and therefore are indicators of the speeds at which galaxies are receding, Hubble's relationship, which has since come to be known as the Hubble law, showed that all galaxies (apart from a few nearby ones which we now know are part of our Local

Group of galaxies) are receding from us with speeds that are proportional to their distances; the greater the distance, the higher the speed of recession. If the Doppler interpretation of the redshifts were correct (which not everyone accepted immediately at the time), Hubble's results clearly implied that the universe is expanding.

The Hubble law and universal expansion

In simple terms, the Hubble law can be written as: velocity of recession (V) = distance (D) multiplied by a constant (H) which gives the numerical relationship between speed and distance and is known as the Hubble constant. Astronomers express the Hubble constant in units of kilometres per second per megaparsec of distance (km/s/Mpc). A megaparsec is a unit of distance equal to a million parsecs, where 1 parsec (see Chapter 1) is equivalent to 3.26 light-years. Hubble's own measured value of this constant (published in 1929 but based on erroneous distance values) was 530 km/s/Mpc – a gross overestimate, as it eventually transpired. Determining the value of the Hubble constant has been one of the most demanding tasks in observational cosmology, mainly due to the sheer level of difficulty associated with measuring the distances of remote galaxies, and it is only in the last decade or so that astronomers have begun to have real confidence in the precision of their results. Current measurements give a value of about 70 km/s/Mpc (less than one-seventh of Hubble's original estimate). This implies, in principle, that a galaxy at a distance of 1 megaparsec should be receding at a speed of 70 kilometres per second, a galaxy a hundred times further away (100 megaparsecs), at 7000 kilometres per second, and so on. In fact, the observed redshifts of galaxies are affected by individual motions induced by the gravitational attraction of neighbouring galaxies, groups, clusters and larger concentrations of matter, which are superimposed on the average underlying expansion rate (which is called the 'Hubble flow').

Strictly speaking, the Hubble constant is not a constant (in the sense of having the same

According to the Hubble law,
the redshift in a galaxy's
spectrum, and the speed
at which it is receding, are
proportional to its distance.
The redshift and recessional
velocity of galaxy B, which is
twice as far away as galaxy
A, are twice as great as
those of galaxy A.

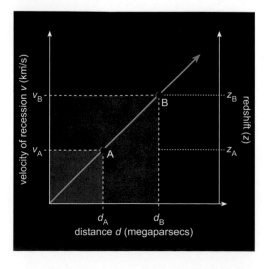

The spiral galaxy NGC 4603, located at a distance of 108 million light-years in the Centaurus cluster, is one of the most distant galaxies in which Cepheid variables have been identified. The locations of several Cepheids in part of the galaxy's spiral arms are identified by little boxes in the right-hand image.

value at all times) at all. The value of H is equal to speed of recession (V) divided by distance (D). If, for example, the galaxies are receding from each other at constant speeds, then by the time the universe has expanded to twice its present size, all distances will have doubled, but their speeds of recession will be the same. The value of the Hubble constant then will be half its present value (same speed divided by double the distance) – 35 km/s/Mpc instead of 70 km/s/Mpc. Because the value of H changes with time, cosmologists use the symbol H_0 to denote the value of the Hubble constant at the present epoch in the history of the universe.

The value of the Hubble constant determines the expansion time, or 'age' of the universe. If the galaxies have been receding at constant speeds since the Big Bang, the time taken for any particular galaxy, travelling at velocity V, to recede to its present distance, D, is just distance divided by velocity (D/V). A galaxy that is ten times further away, has 'travelled' ten times as far, but since the Hubble law tells us that its speed is ten times greater than that of the nearer galaxy, then it will have taken exactly the same time to reach its present distance. The age, or expansion time, of the universe – calculated on the assumption that the galaxies have been flying apart at constant speeds – is known as the Hubble time, and is equal to the reciprocal of the Hubble constant (according to the Hubble law, $V = H \times D$; therefore $H = V/D$; the Hubble time is distance divided by speed and so is equal to D/V, which is $1/H$). For a Hubble constant of 70 km/s/Mpc, the Hubble time is about 14 billion years.

Hubble's original measured value of 530 km/s/Mpc, gave a Hubble time of less than 2 billion years – a figure that was less than the geological age of the Earth! To make matters worse, if the rate of expansion were slowing down, then galaxies would have been moving apart faster in the past than they are now, and would have taken less time to reach their present distances. In that case the actual age, or expansion time, of the universe would be less than the Hubble time, making the conflict between the age of the universe and the ages of its constituent stars and planets even more acute. The current value for the Hubble constant fits in reasonably well with the ages of the oldest known stars, but for much of the twentieth century, the 'age crisis' – the apparent conflict between the age of the universe and the ages of its constituent components – repeatedly

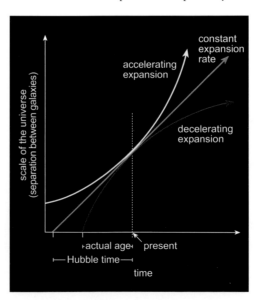

The time that galaxies would have taken to recede to their present distances if they had been receding at constant speeds is known as the Hubble time. If the expansion rate is decelerating, the age of the universe will be less than the Hubble time. If the expansion were accelerating its age would be greater than the Hubble time.

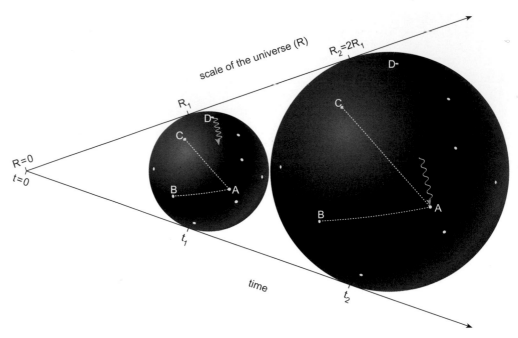

In the 'balloon analogy', galaxies are represented by spots (A, B, C, D, etc.) stuck to the surface of an expanding balloon. As the balloon doubles in size (between times t_1 and t_2), the distance between each galaxy also increases by a factor of two, and no galaxy occupies a unique 'central' position. Each galaxy recedes from every other one with a speed that is proportional to the distance between them (see text). A ray of light departing from galaxy D at t_1 will be stretched to twice its original wavelength by the time (t_2) at which it reaches galaxy A.

reared its head (as we shall see in Chapter 10).

The Hubble law indicates that every galaxy in the universe, apart from our immediate neighbours in the Local Group, is receding from us with a speed proportional to its distance. Although this might seem to suggest that the Milky Way Galaxy is at the centre of the universe, and everything else is rushing away from us in particular, this is an illusion. All the evidence indicates that each galaxy, or each cluster of galaxies, is receding from every other one, and that the whole universe is expanding.

A helpful analogy is to represent the whole of space by the surface of a balloon, and the galaxies by spots stuck to that surface. Ignore the inside and outside of the balloon, and think only of its skin. Suppose that an observer on one of these spots (galaxy A) measures the distance of the other spots (galaxies B, C, D, and so on) and finds, for example, that galaxy C is twice as far away as galaxy B. If the balloon is inflated to twice its previous size, the pattern of spots on its surface will not change, but the separation between each spot will be twice as great as before. The observer on galaxy A will find that galaxy C is still twice as far away as galaxy B, but that the distance through which it has moved is twice as great as the distance through which B has moved. The observer will therefore conclude that each of the

other galaxies is receding with a speed that is proportional to its distance, and will arrive at the Hubble law. However, it is easy to see that galaxy A is not the unique centre of the expansion and that it is the general expansion of the balloon that is causing each galaxy to recede from every other one. Although the balloon itself – a three-dimensional sphere – has a centre, its two-dimensional surface does not. Nor indeed does it have an edge, for a two-dimensional creature – confined to the surface of the balloon and unable even to imagine the third (vertical) dimension – could travel right round its 'universe' without coming to an edge, just as we can circumnavigate the Earth without falling off.

Although our universe most certainly is not the surface of a balloon, it behaves in an analogous way, with each galaxy or cluster receding from every other one, and with no discernable centre or edge. Space may be infinite – in which case it evidently has no boundary – or it may be finite in volume but curved round on itself with no detectable centre or edge; in principle we could circumnavigate such a universe without ever coming to an edge.

The 'balloon analogy' points us towards a different way of looking at the redshifts of galaxies. Instead of thinking of redshifts as a Doppler effect due to the speeds at which galaxies are rushing away through space, cosmologists ascribe redshift to the stretching effect of the expansion of space. The cosmological redshift (denoted by the symbol z) which is observed in the spectra of distant galaxies, relates directly to the amount by which the universe has expanded in the time interval between light being emitted from distant galaxies and arriving here at the Earth. For example, if the separation between the galaxies has doubled in the time it has taken for a light wave to travel from a particular distant galaxy to the Earth (so that we are seeing the galaxy as it was when the universe was half its previous size), the expansion of space will have stretched that wave to twice its original wavelength (the balloon analogy gives a feel for what is going on: if light is represented by waves drawn on the skin of a balloon, then when the balloon is blown up to twice its previous size, all the waves on its surface stretch to twice their previous wavelengths). Consequently, each line in that galaxy's spectrum will have been shifted to twice its normal wavelength, and its measured redshift will be one ($z = 1$).

If we look at a more distant galaxy, so far away that we are seeing it as it was when the separation between galaxies was one-third of present values, its light will have been stretched to three times the original wavelength, and its spectral lines will have been redshifted by a factor of two ($z = 2$). The precise relationship between redshift and the scale of the universe is: $(1 + z) = (R_0/R)$, where (R_0/R) is the factor by which the universe has expanded since the time at which presently observable light was emitted. Putting it the other way round, $1/(1 + z) = R/R_0$, so that a galaxy with redshift 2 is viewed as it was when the universe was $1/(1 + z) = 1/(1 + 2) = 1/3$ of its present size; a galaxy with redshift 3 is seen as it was when the universe was $1/(1 + 3) = \frac{1}{4}$ of its present size, and so on.

Big Bang

If the galaxies are getting further apart now, they must have been closer together in the past, so that if we look far enough back in time, they must at some stage have been exceedingly closely packed. This observation provides the basis for the widely held view that the universe originated a finite time ago by expanding from an exceedingly dense, hot state (perhaps a singularity – a state of infinite compression) in an event that has come to be known as the Big Bang. As early as 1932, the Belgian cleric and mathematical physicist, Abbé Georges Lemâitre, had suggested that the material universe originated from a massive dense concentration of matter – a highly unstable super-atom which contained all the mass of the universe, and which he called the primeval atom. The disintegration of the primeval atom into smaller entities gave rise to the expanding universe that we see today. The modern view of the Big Bang is very different from Lemâitre's, but the foundations of the idea were laid by this talented Jesuit priest.

The key pieces of observational evidence in favour of the Big Bang theory of the origin and evolution of the universe are as follows:

1) The recession of the galaxies: The fact that galaxies and clusters are receding from each other in accordance with the Hubble law is consistent with the idea that the universe is expanding from an earlier compact state – even though of itself this does not constitute absolute proof of a Big Bang.

2) Source counts: By the early 1960s, careful counts of the numbers of galaxies, radio galaxies and quasars to progressively fainter brightness limits, which probed deeper into space and further back in time, showed clearly that the universe was different in the past from today. This fitted in with the idea of a universe that was evolving and changing, as would be the case with a universe expanding from a Big Bang.

3) The age of the universe: The time it has taken for the universe to expand to its present size matches reasonably well with the ages of the oldest known stars, provided that the value of the Hubble constant is not too high (the higher the value of the Hubble constant, the shorter the time it has taken for the galaxies to recede to their present distances and the lower the 'age' of the universe).

4) 'The helium problem': Spectroscopic

observations of planets, stars, gas clouds and galaxies show that hydrogen is by far the most abundant chemical element (making up about 73 percent of the total mass of ordinary matter in the universe) and that helium (the second lightest element) is the second most abundant (nearly 27 percent of the total). The heavier chemical elements such as oxygen, silicon or iron contribute less than 0.1 percent of the total. Astrophysicists believe that those heavier elements were forged in the cores of stars through a succession of nuclear fusion reactions (see Chapter 1) and scattered forth by exploding stars (supernovae) at the end of their lives, to mix in with the hydrogen and helium from which later generations of stars and planets were formed. Whereas this process can account well for the relative proportions of the heavier elements, it cannot account for the observed abundance of the lightest elements – in particular, the large amount of helium. Although helium is continually being produced inside stars by nuclear reactions that convert hydrogen to helium, most of it remains locked up in the cores of living or dead stars. Not nearly enough helium could have been scattered into the surrounding interstellar gas clouds by events such as supernova explosions to account for the quantities that are observed. However, if the universe had been in a sufficiently hot dense state at some stage in its early history, then nuclear reactions would have taken place throughout the whole of the primordial universe, converting hydrogen to helium in the exact proportions that we see today.

5) The cosmic microwave background: The clinching piece of evidence came with the discovery of a weak background of cool microwave radiation, smoothly distributed across the whole sky. In 1964, Arno Penzias and Robert Wilson, physicists working at Bell Labs in New Jersey, USA, were using a sensitive horn-shaped radio antenna to make precise measurements of radio emissions from the Milky Way at a wavelength of a few centimetres. To their surprise, they found that regardless of the direction in which they pointed their antenna to the sky, or the time of day or season

of the year at which they made their observations, they always detected a faint background 'noise' signal. After an exhaustive series of tests and checks to eliminate possible instrumental, terrestrial and atmospheric sources for this excess signal (including removing some nesting pigeons whose droppings might have contributed an unwanted signal!), they concluded that it must be of cosmic origin.

At around the same time, Robert Dicke and Phillip J E Peebles of Princeton University had deduced (as George Gamow had suggested two decades earlier) that if the universe had been in a hot dense state in its distant past, the present-day universe should contain a faint background of radiation – the cooled and feeble remnant of the intense heat of the Big Bang. This primordial radiation would have a distinctive spectrum and a temperature of a few degrees above Absolute Zero. When they heard of Penzias and Wilson's result, they realised that the signal the experimenters had detected might be the predicted afterglow of the Big Bang. Penzias and Wilson published their results in the *Astrophysical Journal* in July 1965. In that same issue, Dicke and his team published a letter outlining their explanation of the phenomenon. Penzias and Wilson's original measurement had been made at a wavelength of 7 centimetres. Subsequent observations made at shorter and longer wavelengths by ground-based, balloon-borne and space-borne instruments (of which more in Chapter 8) have shown beyond all reasonable doubt that the cosmic microwave background radiation is indeed remnant radiation from the Big Bang 'fireball'.

In the face of this array of evidence, and other supporting information, the Big Bang theory has become the firmly established view of the origin and early evolution of the cosmos.

From the Big Bang to the present

The Big Bang was not like an ordinary explosion in which matter erupted from a particular point in a pre-existing empty space. Space, time and matter originated with the Big Bang and 'before' that event there was no space, time or matter in the sense in which we use

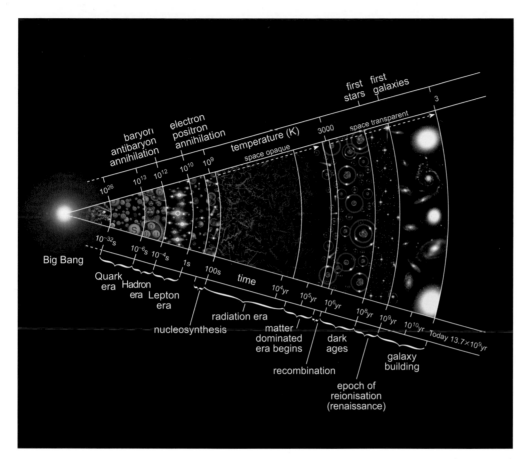

Key stages in the history of the universe from the Big Bang to the present are depicted here in relation to time from the Big Bang and the corresponding temperature of the cosmos. See text for details.

these terms today. Rather than thinking of galaxies as flying away from each other through space, it is better to think of them as being at rest (apart from local individual motions induced by the gravitational pulls of neighbouring clumps of matter) in an expanding space, just like the dots on the surface of the expanding balloon that we used as an analogy for cosmic expansion earlier in this chapter.

Precisely how the universe came into being in that first instant is beyond the scope of present-day physics. However, theoreticians believe that its history after the first microscopic fraction of a second can be described in terms of the known laws of physics. In outline, the story goes as follows:

The universe cooled rapidly during the early stages of its expansion. For example, between 10^{-35} seconds (1 divided by the number 1 followed by 35 zeroes) and 10^{-6} seconds (one millionth of a second) the temperature plummeted from about 10^{27} K (one thousand trillion trillion kelvins) to around 10^{13} K (ten trillion

kelvins). During this extreme high-temperature phase, the universe was filled with intense high-energy radiation. In accordance with Einstein's equivalence between mass and energy ($E = mc^2$) energetic radiation transformed into particle–antiparticle pairs (see Chapter 1) and particles collided with antiparticles, annihilating each other and transforming back into photons of radiant energy. The frequent and rapid collisions between particles, antiparticles and photons maintained a state of equilibrium in which matter and radiation had the same temperature.

In order to make particle–antiparticle pairs of a particular mass, the energies of the photons – which depended on the temperature of the universe – had to exceed a particular threshold. As the universe continued to expand and cool down, photon energies quickly dropped below the threshold at which the more massive particles could be formed. Baryons, such as protons, neutrons and the quarks of which they are constructed, ceased to be formed when the temperature dropped below about 10^{13} K. At this instant – about one microsecond after the beginning of time – quarks combined in groups of three to form protons and neutrons (antiquarks in groups of three to form antiprotons and antineutrons) and quark–antiquark pairs bound themselves together to form short-lived particles called mesons.

The overwhelming majority of baryons and antibaryons then collided and annihilated each other, turning into photons. Had there been an exact equality between the numbers of particles and antiparticles, virtually all the matter in the universe would have been annihilated, and there would be no galaxies, stars, planets or people in the universe today. In fact, there appears to have been marginally more particles than antiparticles. The extent of the imbalance can be gauged by comparing the relative numbers of photons and baryons in the universe today (the photon-baryon ratio). In round figures, there are around a billion photons for every baryon. This implies that when the mutual annihilation was taking place, there were a billion and one particles for every billion antiparticles; each billion antiparticles annihilated a billion particles, leaving one particle as a survivor. The entire baryonic matter content of the universe today is the one in a billion residue of the orgy of self-destruction that took place about a millionth of a second after the beginning of time.

A few seconds later, when the temperature had dropped to around 5 billion kelvins, photon energies dropped below the threshold needed to make electrons and positrons (the positron is the anti-electron), and they in turn mutually annihilated, leaving a one in a billion residue of electrons in a sea of photons, baryons and neutrinos. Mutual interactions between particles continued to transform protons into neutrons and vice versa but, because neutrons are marginally heavier than protons and require more energy for their creation, the proportion of neutrons declined as the universe continued to cool down. By the time the temperature had dropped to around 1 billion kelvins, there were about seven protons for every neutron.

As soon as the temperature had dropped to this level, collisions between neutrons and protons resulted in fusion reactions which created nuclei of deuterium (otherwise known as 'heavy hydrogen', a deuterium nucleus consists of one proton and one neutron). Prior to this, when temperatures were higher, any deuterium nucleus that formed would immediately be blasted apart again by violent collisions with energetic particles or photons. As soon as deuterium began to form, nuclear reactions proceeded rapidly to bind protons and neutrons together to form nuclei of helium-4 (a helium-4 nucleus contains two protons and two neutrons), together with smaller numbers of helium-3 nuclei (each containing two protons and one neutron) and lithium (three protons and four neutrons). Virtually all of the available neutrons were mopped up in this way, the end result being a sea of nuclei comprising 11 hydrogen nuclei (single protons) to each helium nucleus. Because helium nuclei are about four times heavier than hydrogen nuclei, the end result was that about a quarter of the total mass of the ordinary (baryonic) matter in the universe was transformed into helium, leaving about three-quarters in the form of hydrogen.

Big Bang nucleosynthesis – the creation of heavier elements from lighter ones in thermonuclear fusion reactions – effectively came to an end a few hundred seconds after the beginning of time. Thereafter the universe consisted essentially of an expanding, cooling 'soup' of matter and radiation (atomic nuclei of hydrogen, helium and lithium, together with free electrons, photons, neutrinos and – as we shall see later – other, more exotic particles, too) becoming steadily more dilute as it continued to expand. Because photons could travel hardly any distance between collisions with fast-moving electrons or atomic nuclei, space at that time was opaque to electromagnetic radiation. Atoms could not exist because, as soon as a nucleus captured an electron to make a complete atom, the electron would be knocked free again by a collision with another particle or an energetic photon (a process called ionisation).

About 380,000 years after the beginning of time, when the temperature everywhere had dropped to around 3000 K, a dramatic change occurred. Photons no longer had enough energy to ionise (i.e., knock electrons out of) atoms. For the first time in the history of the universe atomic nuclei were able to capture, and hold on to, negatively-charged electrons to make complete, stable atoms. This process is called 'recombination' (although, since it was happening for the first time in the history of the universe, 'combination' might be more appropriate!). The electrons, which were mainly responsible for scattering photons and making the universe opaque to light, were mopped up rapidly so that, in a short time on the cosmic scale, space changed from being opaque to transparent.

Once the sea of free electrons had been mopped up, the primordial photons were able to spread freely through the expanding volume of space. Since then, the expansion of space has cooled and diluted the primordial radiation and stretched (redshifted) the peak wavelength of its spectrum by a factor of more than a thousand. Instead of the universe being filled with visible and infrared photons with a peak wavelength of around 1000 nanometres (0.001 millimetres), which corresponds to radiation with a temperature of around 3000 K, it is now filled with a weak background of microwave radiation, spread across the whole sky. This has a peak wavelength of around a millimetre corresponding to a temperature of 2.725 K, just under three degrees above the Absolute Zero of temperature. The measured temperature and spectrum of this background radiation matches exactly with what theory predicts for a universe which, early in its history, had been a hot dense 'fireball' of matter and radiation.

After recombination, the rapidly diluting radiation from the primeval fireball, stretched by the expansion of space, quickly redshifted beyond the visible range of wavelengths. To human eyes – had any been around at the time to see what was going on – the red glow of the cooling radiation would quickly have faded from view, and the universe would have

become dark. The universe entered a Dark Age from which it would not emerge until the first stars and galaxies began to light up the cosmos.

Precisely how, and when, the first stars and galaxies formed is one of the hottest topics in cosmology and astrophysics and is one we shall return to in more detail in later chapters. However, the basic story is that the seeds for galaxy formation were clumps of matter that, at the time of recombination, were marginally denser than their surroundings. Because of the pull of its own gravity, a clump that was denser than average would expand progressively more slowly than its surroundings; eventually, it would cease to expand, and would then collapse on itself to form a self-contained structure. The favoured paradigm of our time is that galaxies were assembled 'from the bottom up' through a process called hierarchical clustering. Small protogalaxies (precursors of galaxies) formed first and grew through a succession of collisions, mergers and galactic cannibalism (where a more massive galaxy strips matter from, and eventually absorbs, a smaller one) to form successively larger structures. This process is still going on in the universe today.

During the first few hundred million years after the Big Bang, clouds of gas that had been pulled together by gravity transformed themselves into the first generation of stars and, a little later, the first galaxies and quasars. These high-temperature objects emitted intense ultraviolet radiation, which, unlike longer-wavelength light, could not travel far before being absorbed by the sea of hydrogen atoms that filled the universe at that time. When an

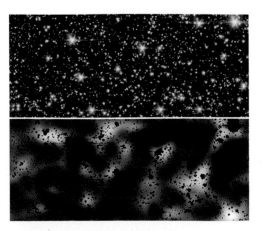

The top panel is an infrared image from NASA's Spitzer Space Telescope of stars and galaxies in the constellation of Draco. The bottom panel is the resulting image after all the stars and galaxies were masked out. The remaining background has been enhanced to reveal what may be the glow of the first stars in the universe.

In this near-collision between two spiral galaxies, gravitational tidal forces generated by the larger and more massive galaxy, NGC 2207 (on the left), have distorted the shape of the smaller one, IC 2163 (on the right), and flung out stars into long streamers stretching a hundred thousand light-years towards the right-hand edge of this Hubble Heritage image. Close encounters, collisions and mergers promote the growth of larger galaxies at the expense of smaller ones.

energetic ultraviolet photon encountered a hydrogen atom, it knocked off an electron, thereby ionising the atom, but was itself destroyed in the process. The first stars and galaxies, though shining brilliantly at ultraviolet wavelengths, were enveloped in, and hidden by, a cosmic fog. Gradually, though, the continual outpouring of ultraviolet photons ionised more and more of the hydrogen, causing bubbles of ionised hydrogen to grow round high-temperature stars and quasars. Eventually, those bubbles became so large that they overlapped completely. By that stage, virtually all of the intergalactic gas (tenuous gas that occupies the space between the galaxies) had become fully ionised once more and, because ionised hydrogen is almost completely transparent to ultraviolet light, the fog that had hidden the early stars and galaxies lifted. By analogy with human history, the period of time when stars and galaxies were emerging from the cosmic fog is known as Cosmic Renaissance. The clearing away of the cosmic fog marked the end of the cosmic Dark Ages.

Since that time, the process of galaxy growth and star formation has continued, bigger galaxies growing at the expense of their smaller neighbours, and galaxies aggregating together to form clusters, superclusters and other large-scale structures such as huge sheets ('walls') and extended filaments, separated by great voids in which few, if any, galaxies are seen.

Cosmic destiny

But what of the future? Will the universe continue to expand forever? If matter and radiation are the sole constituents of the cosmos, then whether or not the universe will continue to expand forever will depend on the mean density of these two ingredients, averaged out over the whole of space.

As the universe expands, the density of matter (the amount of mass in each cubic metre of space) declines in proportion to the cube of the expansion factor (each time the separation between the galaxies doubles, the average density of matter reduces to one-eighth of its previous value). Because the energy contained in photons of radiation is equivalent, through Einstein's relationship ($E = mc^2$), to mass, we can think of the total energy of all the photons in each cubic metre of space as a density, equivalent to the density of matter. Each time the space between galaxies doubles, the average number of photons in each cubic metre of space decreases to one-eighth of the previous number, but in addition, the stretching of space increases the wavelengths of photons and reduces their individual energies. Consequently, the radiation density of the universe declines faster than the density of matter. Although the radiation density far exceeded the matter density in the earliest stages of the history of the universe, it dropped below that of matter about 40,000 years after the beginning of time, and nowadays amounts only to about one three-thousandth of the matter density. If the universe contains nothing else but matter and radiation, the density of matter alone will, to all intents and purposes, determine its future evolution and ultimate fate.

If the mean density exceeds a particular value, known as the critical density, then, in the absence of any other forces, gravity will eventually halt the expansion. The universe will expand to a finite size and then begin to collapse, slowly at first, but ever faster, until all the of the galaxies (or what is left of them by then) merge and the universe ends in a Big Crunch – a state of infinite compression and extreme temperature similar to a Big Bang in reverse. A universe of this kind is called closed. Within such a universe, in accordance with Einstein's general theory of relativity, the overall mean density of matter (and radiation) endows space with a net positive curvature: space curves round on itself in three (or more) dimensions, analogous to the surface of a sphere (see 'The geometry of space', below).

If the mean density is less than critical, gravity will slow the rate of expansion towards a steady value but will never bring it to a halt. The expansion will go on forever. A universe of this kind is said to be open. In an open universe, space is infinite in extent and its overall curvature is negative: at every location in space, the geometry of space is analogous to the shape of a saddle. Parallel lines drawn at a particular location on a negatively-curved surface eventually diverge from each other.

The same holds true in a negatively-curved three-dimensional space or four-dimensional space-time.

If the density is precisely equal to the critical value, the universe is just, but only just, capable of expanding forever. The recessional velocities of the galaxies will decrease, approaching ever closer to zero, but not becoming precisely zero until an infinite amount of time has passed. Such a universe is called flat because, on the large scale (ignoring the localised curvatures caused by concentrations of mass such as galaxies and clusters) the net curvature of space is zero. In such a universe, space is infinite in extent, and the ordinary laws of Euclidean geometry apply: as is the case for a flat sheet of paper, in a flat universe, the shortest distance between two points is a straight line, and parallel lines never meet.

By common convention, cosmologists denote the ratio of the actual mean density to the critical density by the Greek character, Omega (Ω). If the mean density is greater than the critical density, then Omega is greater than 1 ($\Omega > 1$), whereas if the mean density is less than critical, Omega is less than 1 ($\Omega < 1$). If the mean density is precisely equal to the critical density, then Omega = 1. If the expansion of the universe is controlled solely by the gravitational influence of matter and radiation, then space will be closed and positively-curved if $\Omega > 1$, but open and negatively-curved if $\Omega < 1$. If $\Omega = 1$ the universe is flat and 'sits on the fence' between the open and closed models.

The geometry of space

On a flat surface, the shortest distance between two points is a straight line. To construct a pair of parallel lines, you can draw any straight line and then two other lines that intersect the first line at different points but at the same angle. These lines will never meet, and will always be separated by the same distance. On the surface of a sphere, the shortest distance between two points is a curved line, and 'parallel' lines eventually intersect. For example, meridians (lines of longitude) all cross the equator at right angles and, because they make the same angle with the equator we can think of them as parallel; however, as they extend further north and south of the equator, they come progressively closer together, eventually intersecting at the north and south poles. In a positively-curved space, the shortest distance between two points is a curved line, and 'parallel' lines drawn at any particular location in space eventually meet. In an analogous way, 'parallel' lines in a positively-curved three-dimensional space, or in four-dimensional space-time, eventually intersect. Conversely, parallel lines drawn on a negatively-curved, saddle-shaped surface diverge.

The diagram on the right shows the behaviour of 'parallel' rays of light in closed, open and flat spaces.

Each of these possibilities corresponds to one of the range of theoretical models of the universe derived in the early 1920s by Russian mathematician, Alexandr Friedmann, well before the recession of the galaxies and the expansion of the universe had been discovered by observational astronomers. The open, flat and closed models are all known as Friedmann models. The flat model is also known as the Einstein–de Sitter universe, because it corresponds to a model which Einstein developed, in association with Dutch astronomer Willem de Sitter in 1931, based on the assumption that space was filled with a uniform distribution of matter alone.

The actual value of the critical density depends on the value of the Hubble constant. For example, if (as current observations indicate), the value of the Hubble constant is about 70 km/s/Mpc, the critical density works out to be about 10^{-26} kg/m³, which is about a hundred trillion trillion times less dense than the air that you are breathing as you read this sentence. The critical density is equivalent to there being, on average, between five and six hydrogen atoms in every cubic metre of space, averaged over the whole universe.

In principle, astronomers ought to be able to determine whether the universe is open, flat or closed, and settle the question of whether

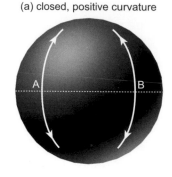

(a) closed, positive curvature

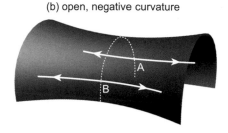

(b) open, negative curvature

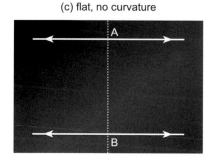

(c) flat, no curvature

A depiction of how the scale of space, or the separation between galaxies, changes with time in open (ever-expanding), flat, and closed (ultimately-collapsing) universes. The symbol Ω (Omega) denotes the ratio of the actual mean density of the universe to the critical density (the mean density of a flat universe).

open (ever expanding) $\Omega < 1$
flat $\Omega = 1$
closed (ultimately collapsing) $\Omega > 1$
scale of the universe (separation between galaxies)
Big Bang present Big Crunch
time

or not it will expand forever, by measuring its mean density, or some equivalent quantity, such as the overall curvature of space, or the rate at which the expansion is slowing down. In practice, making precise measurements of any of these quantities is an extremely difficult and challenging task – one that has preoccupied observational cosmologists down the decades since the discovery of the recession of the galaxies. One thing is clear, however: there is not nearly enough luminous matter – the stuff we can see directly through telescopes – to add up to anything approaching the critical density. But, as the next chapter will show, luminous matter is only a small part of the story.

More than Meets the Eye

If luminous matter were its sole constituent, astronomers would be able to determine the mean density of the universe by selecting a volume of space large enough to be a fair representative sample of the cosmos as a whole, adding up the masses of the visible stars and galaxies in that volume, and then dividing by the volume. When astronomers make measurements of this kind, they find that luminous matter – the stars and glowing gas clouds that delineate the visible galaxies – contributes only about half of one percent of the critical density (see 'Visible light and the density of the universe' on page 34). If the universe contained nothing else but luminous matter, gravity would be far too feeble ever to halt its headlong expansion. The universe would be open, and destined to expand forever.

However, the universe contains a great deal more than directly meets the eye. There is abundant and compelling evidence to show that galaxies, clusters and the universe as a whole contain far more dark than visible matter. Dark matter does not radiate detectable quantities of visible light or any other kind of radiation. It does not glow, nor does it reveal its presence by absorbing light from background stars, galaxies or glowing gas clouds. This 'dark stuff', whatever it may be, plays a vital role in determining the density, geometry, evolution and fate of the universe.

The primary evidence for the existence of large amounts of dark matter is provided by studies of the way in which spiral galaxies rotate and by analyses of the internal motions of galaxy clusters. Although the first strong evidence for the presence of large amounts of dark matter emerged, as early as 1933, from studies of galaxy clusters, let's look first at spiral galaxies.

Dark matter in spiral galaxies

In a spiral or disc-shaped galaxy, such as our own Milky Way system, stars or gas clouds in the flattened galactic disc revolve in near-circular orbits around the galactic centre. The orbital speed of a star depends on the net gravitational pull of all of the mass that is contained inside its orbit. Provided that mass is distributed throughout the galaxy in a reasonably symmetrical way, matter that lies outside

M51, the Whirlpool galaxy, is a classic spiral galaxy with two well-defined dominant spiral arms highlighted by hot young stars and glowing gas clouds. However, spiral galaxies such as this one contain far more dark matter than luminous matter.

Visible light and the density of the universe

One way of calculating the mass of a galaxy would be to measure the total amount of light that it emits, divide by the average amount of light that a single star emits and then multiply by the mass of an average star. To do this, we need to know the mass-to-light ratio for the population of stars of which that galaxy is composed (in effect, the ratio of the average mass of those stars to their average luminosity; or, to put it another way, how much mass is required to generate a given light output). Astronomers find it very convenient to express the mass and luminosity of stars using a scale on which mass of the Sun (the solar mass) = 1 and the light output of the Sun (the solar luminosity) = 1. If each and every star in a galaxy were identical to the Sun, the mass-to-light ratio (M/L) would be 1 (M/L = 1/1 = 1). In a galaxy such as the Milky Way, however, most stars are less luminous than the Sun, so the mass-to-light ratio is greater than 1 (the figure for the Milky Way Galaxy is about 6). It turns out that the average mass-to-light ratio for a population of stars is a number somewhere between 1 and 10. At one end of the scale, in a galaxy where a lot of star formation is taking place, there will be a large number of highly luminous hot young stars, and the average M/L will be towards the bottom end of the range. By contrast, in a galaxy which has long since converted practically all of its available hydrogen gas into stars, the short-lived highly luminous stars will have faded away; most of its light will be provided by cooler, dimmer stars, and its mass-to-light ratio will be near the upper end of the range.

When astronomers make measurements of this kind, they find that, on average, the combined light output of all the visible objects in a cubic megaparsec of space (a box measuring 1,000,000 parsecs – 3,260,000 light-years – along each side) is equivalent to the luminosity of about a hundred million Suns. If we multiply the luminosity in the box (10^8 solar luminosities) by the average mass-to-light ratio (the amount of mass needed to generate that amount of light) – say about 5 solar masses – we can calculate the mass of all the stars in a typical 1 megaparsec cube (in this example, it works out at 5×10^8 solar masses). If we now multiply this figure by the mass of the Sun (2×10^{30} kilograms) we get the number of kilograms of luminous material in a cubic megaparsec (which is equivalent to about 3×10^{67} m³). Divide the total mass by the volume, and we get a mean density for luminous matter. When we compare that figure to the critical density, we find that luminous matter adds up to only about half of one percent of the critical density. Putting it another way, Ω (for luminous matter) is in the region of 0.005.

the star's orbit will have little or no effect on the orbital motion of the star. For example, if a star's orbit were surrounded by a spherical shell of matter, different bits of the shell would 'pull' in different directions; the gravitational attraction in any particular direction would be cancelled by the attraction in the opposite direction, and the net effect on the star's motion would be zero. The net gravitational pull on the star will be the same as if all of the mass inside its orbit were concentrated in a single body at the centre of the galaxy.

When a small body travels in a circular orbit around a much more massive one, its speed depends on the mass of the central body and the radius of the orbit: the more massive the central body, the higher the orbital speed; the larger the orbital radius, the lower the orbital speed. Specifically, the orbital speed (or circular velocity) is inversely proportional

to the square root of orbital radius so that, for example, if one orbiting body is four times as far from the central mass as another, its orbital speed will be half that of the inner one ($1/\sqrt{4} = \frac{1}{2}$). Within the Solar System, where the Sun is by far the most massive body, the orbital speeds of the planets obey this relationship. Orbital motion which obeys this relationship is known as Keplerian motion, after the German astronomer Johannes Kepler who, in the early seventeenth century, was the first to uncover the laws that govern planetary motion.

The Sun revolves around the centre of the Milky Way Galaxy in a near-circular orbit in much the same sort of way as a planet revolves around the Sun. Located at a distance of about 27,000 light-years from the galactic centre, the Sun has an orbital speed of 230 kilometres per second, and takes some 225 million years to complete one circuit around the Galaxy.

of these, and other, nearby galaxies, and the streams and blobs of material that have been torn from them, give further clues to the overall mass and extent of the Milky Way Galaxy and its dark halo.

Another clue comes from the fact that the Milky Way Galaxy and the Andromeda galaxy (a spiral galaxy similar to but somewhat larger than our own) are approaching each other at a speed of around 120 kilometres per second (they are expected to collide and merge in about 3 billion years' time). At the time of their birth, these galaxies would have been moving apart because of the general expansion of the universe. Consequently, astronomers can obtain a rough estimate of their combined mass by working out how much mass would have been needed to slow down and turn round their initial recession and cause them to fall together as they are now doing.

Putting all the available evidence together, it seems likely that the dark halo of the Milky Way extends out to at least 500,000 light-years, and that the total mass of our Galaxy is in the region of 1000 billion to 1500 billion solar masses. By comparison, its total luminous mass adds up to only about 120 billion solar masses. It appears that our Galaxy contains at least ten times as much dark matter as luminous matter.

Dark haloes around elliptical galaxies

There is evidence, too, that many (though not necessarily all) elliptical galaxies are surrounded by dark matter haloes, but the evidence is harder to acquire and interpret. Unlike spirals, ellipticals do not have a clearcut pattern of rotation in a single plane from which a rotation curve can be constructed; instead, their constituent stars orbit in all kinds of random directions. However, the motion of each individual star is still determined by the amount of mass that lies inside its orbit. Depending on the orientation of its orbit, part of a star's velocity will be directed towards or away from us, and this will give rise to a Doppler shift in its spectral lines – a blueshift if it is approaching or a redshift if it is receding. This component of its motion, and hence the amount of redshift or blueshift in its spectrum,

will be highest if the star happens to be heading directly towards or away from us, but zero if it is moving at right angles to our line of sight (neither approaching nor receding). The spectrum of the galaxy as a whole (or of a particular region within the galaxy) consists of the combined light of vast numbers of stars, some approaching and some receding, with a wide range of velocities and a correspondingly wide range of redshifts and blueshifts in their individual spectra. Consequently, each line in that galaxy's spectrum is smeared out over a range of wavelengths – the greater the range of radial velocities, the broader the line. By measuring the widths of an elliptical galaxy's spectral lines, astronomers can work out how fast its constituent stars are moving, and hence measure the amount of mass that is contained within its visible extremities.

Measurements of the broadening of spectral lines at different distances from the centre of elliptical galaxies show that the velocities of stars remain fairly constant with increasing distance, which suggests that, as with spiral

At least ten times as massive as the Milky Way Galaxy, the giant elliptical galaxy, M87 is surrounded by about ten thousand globular clusters, many of which are seen as tiny point-like objects in this image. Studies of the motions of these clusters and of x-ray emission from hot gas that surrounds this galaxy indicate it is surrounded by a halo of dark matter at least ten times as massive as its visible population of stars. Image courtesy of Canada-France-Hawaii Telescope/J-C Cuillandre/Coelum.

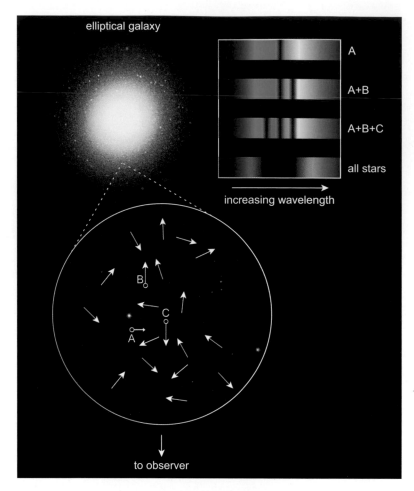

elliptical galaxy

A

A+B

A+B+C

all stars

increasing wavelength

to observer

The spectrum of a region within an elliptical galaxy is composed of the combined spectra of a vast number of stars. The series of spectra shown from top to bottom on the right, show a line in the spectrum of a single star (A); the combined spectra of two stars (A and B), one of which (B) is receding; the combined spectra of three stars (A, B and C – which is approaching); the combined spectra of all of the stars where all of the lines merge into a single broad one. The effect has been exaggerated here for clarity.

the European Space Agency's XMM-Newton, have revealed how the temperature of this hot gas varies with distance from the centre, and have shown that, even at large distances, the gas is so hot that it would readily have escaped from its host galaxy unless held in by a gravitational attraction considerably stronger than its visible stars can provide.

Observations of this kind have shown that many of the more massive giant ellipticals, such as M87 in the Virgo cluster, are embedded within dark matter haloes that typically are five to ten times as massive as the total mass of the stars that they contain. The situation is more diverse, and less clear-cut, with smaller, less massive, ellipticals.

Because most of the ellipticals with x-ray haloes are embedded within groups and clusters which themselves (as we shall see later) have extended haloes of hot intracluster gas (ionised gas that occupies the space between a cluster's member galaxies), it is difficult to determine whether the x-ray emitting material belongs to a particular galaxy or to the cluster as a whole. For that reason, some astronomers have undertaken searches for isolated elliptical galaxies with x-ray haloes. One such example is NGC 4555, a fairly large elliptical galaxy which was investigated by Ewan O'Sullivan, of the Harvard-Smithsonian Center for Astrophysics in Cambridge, Massachusetts and Trevor Ponman, of the University of Birmingham, UK. They found it to be surrounded by an envelope of ionised gas some 400,000 light-years in diameter. By comparing the total amount of mass needed to hold on to this hot gas cloud with the total amount of light emitted by its constituent stars, they found that this particular galaxy contains at least five times as much dark matter as luminous matter; just like the giant ellipticals (such as M87) which inhabit the more crowded environments of galaxy clusters.

Other probes of elliptical galaxy haloes include the distribution and motion of globular clusters, satellite galaxies, and other neighbouring galaxies. But although measurements of these objects provide information about the overall mass of a galaxy's halo, they provide only limited information about the

galaxies, a substantial amount of mass may lie beyond their visible boundaries. But, because ellipticals, unlike spirals, do not possess radio-emitting clouds of hydrogen gas that can be used as tracers of orbital motion at large distances from their centres, it is difficult to get a handle on their total masses. Some, however, are embedded within extensive, though tenuous envelopes of ionised gas which, with temperatures in the region of ten million to a hundred million degrees (10^7–10^8 K), are so hot that they emit copious quantities of x-rays. Temperature is a measure of the speeds at which ions and electrons are rushing around within these clouds – the higher the temperature, the higher their speeds, and the greater the gravitational pull that is required in order to prevent the gas cloud from dissipating. Detailed high resolution observations of these hot envelopes, made by orbiting spacecraft such as NASA's Chandra X-ray Observatory and

way in which mass varies with distance from the galaxy's centre. Recently, however, astronomers have begun to use planetary nebulae as tracers of orbital motion in elliptical galaxies. A planetary nebula is a shell of gas which is expelled by a dying star at a late stage in its life cycle. Because most stars are expected to reach this stage eventually, the distribution of planetary nebulae in a galaxy should tie in closely to the distribution of ordinary stars, and their motions should be representative of the overall population of stars. Planetary nebulae can be identified because their emission lines (bright lines of light at particular wavelengths which are emitted by glowing gas) are easier to identify and measure than the dark lines produced by ordinary stars.

In 2003, Aaron J Romanowsky (then at the University of Nottingham, UK) and co-workers used a specialised Planetary Nebula Spectrograph to obtain the velocities of hundreds of planetary nebulae in a number of nearby elliptical galaxies. They found that the spread of orbital velocities increases with increasing distance from the centres of two of the most massive ellipticals in the Virgo cluster, which fits in with these galaxies possessing massive dark matter haloes (as x-ray observations and studies of globular clusters had already shown). However, with several of the more modest elliptical galaxies that they investigated, they found that the velocities of planetary nebulae decline with increasing dis-

tance, as if there were little or no dark matter present at all.

One possible explanation for this effect is that stars in the outer parts of elliptical galaxies move in random directions along paths which may be anything from circular to exceedingly elongated ellipses. In the latter case, when we look towards the outer edges of elliptical galaxies, many of the stars would be moving across the line of sight, rather than towards or away from us, and this would reduce their perceived average line of sight velocities. This would lead astronomers to underestimate the amount of dark matter in the outer parts of those galaxies. This possibility fits in well with the currently favoured theory of galaxy formation (see Chapter 5), which suggests that elliptical

The image on the left, which was obtained by the Chandra X-ray Observatory, shows that the large, isolated elliptical galaxy, NGC 4555, is embedded in an extensive cloud of gas with a temperature of about 10 million kelvins. The hot gas cloud has a diameter of about 400,000 light-years, roughly twice that of the visible galaxy (right).

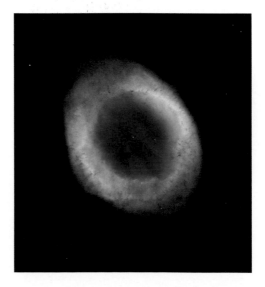

The Ring Nebula (M57) is a planetary nebula, a shell of gas that was ejected several thousand years ago by a dying star. The gas glows because it is bathed in intense ultraviolet light from the faint, but extremely hot, remnant star which is located at the centre of the nebula.

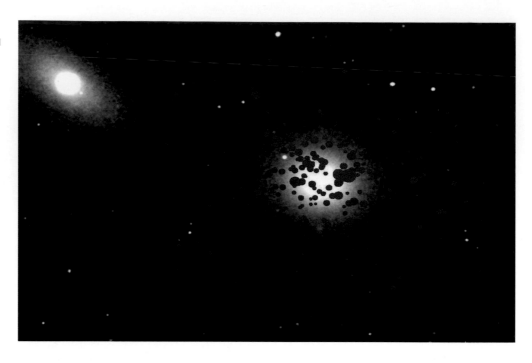

The dots show the positions of planetary nebulae located in and around the elliptical galaxy, M105. The colour of each dot shows whether the nebula is approaching or receding, and its size indicates its speed. The image shows that planetary nebulae can be detected well beyond the apparent luminous boundary of the galaxy, and that the dots tend to get smaller far from the galaxy. This implies low speeds and a lack of dark matter. Courtesy of the PN.S Consortium.

galaxies are formed when disc (spiral) galaxies merge together. A recent numerical simulation, which was published in 2005 by Avishai Dekel, of the Hebrew University in Jerusalem, and co-workers, shows that during mergers of this kind gravitational tidal forces would indeed fling stars into highly elongated trajectories.

Another possibility is that the dark matter haloes surrounding these particular elliptical galaxies may be so much larger than the vis-ible galaxies that the influence of dark matter only becomes dominant at distances greater than the limits to which matter has so far been observed directly. Or, again, perhaps these galaxies have been stripped of their dark matter haloes by close encounters with other galaxies. Nevertheless, the detection of several ellipticals which apparently have little or no dark matter, and which are relatively faint at x-ray wavelengths (and therefore may not have enough mass to be able to retain hot gas), does pose problems for the standard dark matter halo hypothesis. Whether or not a galaxy has an extensive dark matter halo may depend on how it was formed and on the environment within which it is located.

These anomalies notwithstanding, it looks on balance as if spiral galaxies and many, although not necessarily all, elliptical galaxies are embedded within extensive dark matter haloes that outweigh their luminous components of stars and gas by a factor in the region of 5–10, and sometimes considerably more.

Dark galaxies?

The universe may also contain galaxies that are completely dark – conglomerations of matter which emit no detectable light because they contain few, if any, stars. A particularly

Dark Galaxy VIRGOHI 21 has no starlight but radio waves from neutral hydrogen betray its existence. The contours superimposed on this optical image indicate how much gas was detected.

striking example of what may be an object of this kind was reported in 2005 by a team of astronomers headed by Robert Minchin of the University of Cardiff, Wales. The object in question, named VIRGOHI21, is located in the Virgo cluster of galaxies, some 50 million light-years away. Its presence was revealed through the detection of the distinctive 21-centimetre wavelength radio emission that characterises clouds of atomic hydrogen gas. The radio observations, which were carried out using the Lovell Telescope at Jodrell Bank, England, (and subsequently confirmed by more detailed observations made by the Westerbork array of radio dishes in the Netherlands) revealed an extended distribution of hydrogen gas with a total mass equivalent to about 100 million suns, spread over a region of space about 50,000 light-years across. The observers found that the gas is moving just as if it were swirling around the core of a normal star-filled galaxy with a mass of around 10 billion solar masses (10 percent of the mass of the Milky Way Galaxy) – a mass which is about a hundred times greater than the mass of the hydrogen cloud itself.

An ordinary galaxy with that sort of mass, at that sort of distance, would have been bright enough to have been seen quite easily with a modest-sized telescope, but when the team used the 2.5-metre Isaac Newton Telescope on La Palma, in the Canary Isles, to search for any hints of visible light in that region of the sky, they found nothing. In light of the evidence, the most likely explanation is that VIRGOHI21 is made almost entirely of dark matter. It appears to be a dark matter halo that does not contain the expected luminous galaxy – a system which has not converted any of its constituent gas into stars. The existence of dark galaxies is predicted by the favoured theory of galaxy formation (see Chapter 6). VIRGOHI21 appears to be the first well-attested case of an object of this kind.

Dark matter in clusters of galaxies

Evidence for the existence of large amounts of dark matter in galaxy clusters was first uncovered by Swiss-born astronomer, Fritz Zwicky

as long ago as 1933. Observing from Mount Wilson, California, Zwicky made a detailed study of the Coma cluster of galaxies, which lies at a distance of about 300 million light-years and is estimated to contain about two thousand members. By measuring the range of Doppler shifts in the spectra of individual member galaxies, he was able to calculate the spread of galaxy velocities inside the cluster (the velocity of recession of the cluster as a whole is given by the average motion of its members). Because the motion of individual galaxies within the cluster to which they belong depends on the gravitational pull of their immediate neighbours and on the overall gravitational field of the cluster as a whole, Zwicky was able to use his measurements to 'weigh' the Coma cluster. What he found was that the total mass of the cluster was vastly greater than the amount of visible matter that it contained.

When he divided the total mass of the cluster by the number of galaxies within it, he arrived at a figure of about 50 billion solar masses per galaxy. But the total light output of the average Coma galaxy was only about 90 million times the luminosity of the Sun. Taken together, these figures showed that the ratio of mass to luminosity (on a scale in which the luminosity of the Sun = 1 and the mass of the Sun = 1) was more than 500:1, which implied that the cluster contained 500 times as much mass as would be the case if its component galaxies consisted entirely of stars identical to the Sun. It was known at that time that most stars are fainter than the Sun, and the stars themselves have a mean mass-to-light ratio of about 10 (the mean mass-to-light ratio for the mix of stars in galaxies ranges from 1 to 10, depending on the type of galaxy). Even so, Zwicky's fairly crude calculation indicated that the Coma cluster contained about fifty times as much mass as he would have expected, based on the light output of its member galaxies. Faced by these extraordinary numbers, Zwicky concluded that the Coma cluster must contain a huge amount of exceedingly dim, or completely dark, matter – a great deal more mass than meets the eye. A few years later, in 1936,

The Coma cluster is a typical rich cluster, which has more than a thousand member galaxies. This composite of images, obtained using the Isaac Newton Telescope on La Palma, shows that its central region is dominated by two giant elliptical galaxies. As is usual for clusters of this richness, the great majority of galaxies are elliptical, with only a few spirals.

The contours on this image of the very massive distant cluster of galaxies, RXCJ1206.2-0848 show the brightness distribution of x-rays radiated by the extensive cloud of hot intracluster gas that pervades the space between its member galaxies.

Sinclair Smith obtained similar results for the nearby Virgo cluster.

Although more recent estimates put the mass-to-light ratio for the Coma cluster at more like 200, this is still extremely high compared to the mass-to-light ratio for the population of stars in a typical galaxy. Since Zwicky made these pioneering observations, astronomers have found a similar excess of gravitational mass to luminous mass – dark matter to visible matter – in all other galaxy groups and clusters, with the typical mass-to-light ratio for a cluster being about 300. These figure imply that clus-

ters contain up to a hundred times as much mass as is directly visible in the form of the stars which dwell in their constituent galaxies.

X-ray observations show that many clusters contain large amounts of exceedingly hot ionised gas, which, because it occupies the space between the member galaxies of clusters, is known as intracluster gas. Almost certainly, the reason why this gas is so hot is that it has been heated by energy that was released when its constituent atoms, nuclei and electrons were pulled inwards by the combined gravitational influence of the cluster's dark matter halo and its constituent galaxies. Particles pick up speed while falling from 'a great height'; the faster they move, the more violently they collide, and the higher their temperature becomes. Eventually, when the pressure exerted by the fast-moving gas particles becomes high enough to counterbalance the inward pull of gravity, the cloud reaches a state of equilibrium and ceases to contract.

Because the amount of x-radiation emitted (per second) by the cluster as a whole depends on the total amount of hot gas that it contains, astronomers can use x-ray observations to

The Sunyaev-Zeldovich effect

Further data on the amount and distribution of ionised gas in clusters comes from a phenomenon called the Sunyaev-Zeldovich (or 'S-Z') effect, the essence of which is that clouds of high-temperature ionised intracluster gas leave their imprint on the cosmic microwave background. When low energy microwave photons scatter off (rebound from) fast-moving electrons in the intracluster gas, they pick up energy, and are blueshifted to shorter wavelengths (the higher the energy of a photon, the shorter its wavelength). This displaces the whole spectrum of the microwave background, in that region of the sky, to shorter wavelengths and therefore reduces its brightness at the longer wavelengths that are typical of the microwave background as a whole. The effect is small – the dip in temperature is less than a thousandth of a degree – but measurable. Because the magnitude of the effect depends on the number of electrons along the line of sight through the cloud, and on its temperature, astronomers can use the S-Z effect to measure the amount of gas in the cluster.

'weigh' the intracluster gas. Furthermore, by measuring the energies of the x-rays at different distances from the centre of the cloud astronomers can work out how its temperature varies with increasing distance from its centre and, in particular, can find out the temperature towards its outer fringes. If they assume that the gas cloud is in a state of equilibrium, they know that the inward pull of gravity must be enough to prevent the intracluster gas from dispersing, and can therefore work out the total amount of mass that is pulling on the gas. Observations of this kind show that x-ray clusters typically contain about six times as much mass in the form of hot ionised gas as they do in the form of starry galaxies, but that even when the masses of its luminous galaxies and intracluster gas are added together, their combined mass still adds up to little more than a tenth of the overall mass of the cluster.

Through gravity's lens

Further evidence for the existence of dark matter comes from the phenomenon of gravitational lensing. According to Einstein's general theory of relativity (see Chapter 1) the presence of mass (or energy) distorts, or 'curves' space so that a ray of light passing close to a massive body will be deflected by an amount that depends on the mass of the body and the minimum distance between the body and the ray. Consequently, as Einstein first proved in 1936, a massive foreground object can act as a lens to produce a magnified image of a background object. If the background source (for example, a star), foreground object and observer are exactly lined up, the rays of light bending round the massive foreground object will be spread out into a ring of light, called an Einstein ring. If the alignment is imperfect, the background object will appear as two or more distorted arc-shaped images. The first gravitational lens to be identified was the so-called double quasar 0957 + 561 (a designation which is based on its coordinates in the sky), two distorted images of the same background quasar which were detected in 1979 by radio astronomers Dennis Walsh, Robert Carswell and Ray Weymann.

A foreground distribution of mass (or energy), such as a cluster of galaxies, will also act as a lens. The deflections that occur when rays from a distant background galaxy, or cluster of galaxies, pass through a foreground cluster cause light rays to converge to form distorted, arc-shaped images of the background objects which are larger and brighter than those objects would appear to be if the 'lens' were not present. In this way, astronomers have been able to study galaxies which are too far away to be seen directly. A classic example of this phenomenon is the rich cluster Abell 2218, which has produced more than 120 arc-shaped images of galaxies which themselves are members of a much more distant cluster, five to ten times as far away as Abell 2218 itself. As expected from theory, the observed arcs are lined up at right angles (i.e. tangentially) to the direction from each galaxy to the centre of mass of the lensing cluster. By analysing the deflections that have taken place, astronomers can map the distribution of matter – both

The gravitational field of a massive foreground galaxy can act like a lens to produce two images of a distant background object, such as a quasar.

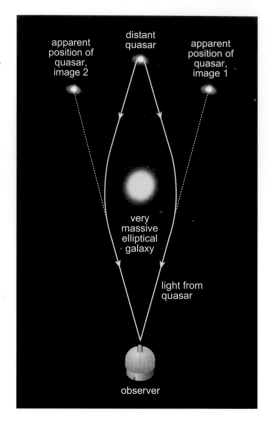

luminous and dark – inside the lensing cluster, and work out its total mass. Intriguingly, Fritz Zwicky (again!) was the first to suggest, back in 1937, that gravitational lensing might provide a way of measuring the mass of galaxy clusters. Studies of the lensing effects produced by galaxy clusters confirm that clusters of galaxies contain from 10 to 100 times as much dark

matter as luminous matter.

The general distribution of dark matter in the universe exerts a more subtle effect known as weak lensing. Rays of light that are travelling towards us from remote galaxies follow tortuous paths through the cosmos because they are deflected to some degree by every clump of matter that they meet along the way. Every bend in the light path stretches the image of a background elliptical galaxy in one direction or another, so that its apparent shape becomes distorted by an amount that depends on the distribution of matter along the line of sight. Whereas the true shape of any individual elliptical galaxy is not known, astronomers can use statistical analysis of the average distortion of large numbers of galaxies to measure the overall distribution of dark matter in the universe. Studies of this kind first began to bear fruit in the year 2000. In the spring of that year no less than four different research groups independently, and more or less simultaneously, succeeded in measuring and analysing minor distortions in the images of between 100,000 and 200,000 distant galaxies. These measurements had been obtained by wide-angle CCD cameras (electronic devices that produce digital images) on telescopes located on Mauna Kea, Hawaii, in the Chilean Andes, and at La Palma, in the Canary Islands. This research produced the first ever 'maps' of the large-scale distribution of dark matter in a selected region of the

This image of the rich cluster of galaxies, Abell 2218, shows a spectacular example of gravitational lensing. The numerous thin arcs are distorted images of a much more distant population of galaxies produced by the focussing effect of the gravitational field of Abell 2218.

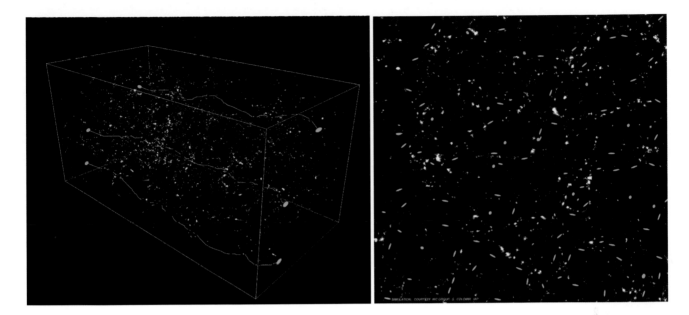

sky. With improving instrumentation and techniques, weak lensing is now developing into one of the most powerful methods of mapping the distribution of matter in the universe (see Chapter 12). Weak lensing confirms that there is insufficient matter (luminous or dark) to halt the headlong expansion of the universe.

Further clues to the amount of matter – both luminous and dark – come from measurements of large-scale flows. Superimposed on the smooth overall expansion of the universe in accordance with the Hubble law (the so-called 'Hubble flow') are additional motions caused by the gravitational influence of massive concentrations of material. By mapping these 'extra' large-scale motions, and relating them to the observed distribution of galaxies, astronomers can work out how much matter density is needed to explain these large-scale flows.

Large amounts of dark matter seem to be essential in order to explain how galaxies form at all. The cosmic microwave background, alluded to in Chapter 2, contains a multitude of marginally hotter and cooler patches (called temperature fluctuations) which reveal the presence of underlying differences in the density of ordinary matter (called density fluctuations). These fluctuations existed at the time

(Left) This simulation shows the web-like distribution of dark matter in a large volume of space (a billion light-year cube) and the deflections experienced by rays of light (yellow) travelling from three distant galaxies, represented by three blue discs at the rear (left) of the cube. The net effect of these deflections is to elongate each galaxy's image very slightly.
(Right) This view shows what an observer at the front of the box would see when looking at the galaxies in the sky. On average, the galaxies are stretched along directions parallel to the intervening filaments of dark matter. Courtesy of S. Colombi (IAP), Canada-France-Hawaii Telescope Team.

Galaxy surveys

In recent years, surveys of the positions and redshifts of hundreds of thousands of galaxies have begun to provide a wealth of information about the large-scale distribution of luminous and dark matter. Pre-eminent among these are the Two-Degree Field Galaxy Redshift Survey (2dFGRS), and the ongoing Sloan Digital Sky Survey (SDSS). The 2dFGRS was carried out by the 3.8-metre Anglo-Australian Telescope, in Australia, with the aid of a device that enabled astronomers to obtain spectra (and hence redshifts) of large numbers of galaxies simultaneously (the '2dF' referred to the fact that the instrument could image an area of sky just over two degrees in diameter - much larger than the field of view of a conventional large telescope). By the time the survey concluded, in 2003, it had accumulated data on more than 220,000 galaxies. The ongoing Sloan Digital Sky Survey (SDSS), which operates from the Apache Point Observatory in New Mexico, images galaxies and their spectra electronically. By the time its initial five year survey has been completed in 2006, it is expected to have acquired similar data for more than a million galaxies and a hundred thousand quasars. Both projects have produced detailed three-dimensional maps of the distribution of galaxies out to distances in the region of 2-3 billion light-years which vividly reveal the frothy distribution of galaxies, clusters, superclusters, filaments, sheets and voids.

In the Sloan Digital Sky Survey, galaxies are identified in two-dimensional images (right), then have their distances determined from their spectra to create a three dimensional map (left) of the distribution of galaxies in which each galaxy is shown as a single point

– around 380,000 years after the Big Bang – when the microwave background radiation was released and when clumps of ordinary matter first became free to begin to pull themselves together, under the action of gravity, to form galaxies, clusters and larger-scale structures in the universe. However, as will be explored later in Chapters 5 and 8, the temperature differences – and hence the underlying density differences – were so slight that there would not have been enough time for them to form the multitude of galaxies and clusters that we see in the universe today, without the assistance of the additional, dominant, gravitational influence of dark matter.

There is a huge accumulated body of evidence in favour of dark matter, but the dark matter scenario is not without its problems. For example, as studies of planetary nebulae embedded within at least some of the more modest elliptical galaxies have shown, some ellipticals appear to contain very little or no dark matter. Again, standard theories predict there should be a sharp rise in density (a density 'spike') towards the centres of dark matter haloes (i.e. in the cores of galaxies), but this effect has been very marginally detected in only a very few cases. Furthermore, standard models of galaxy formation through the merging of dark matter haloes imply that large galaxies should be surrounded by many more satellite systems and small dark companions than observations so far have revealed. These and other issues have encouraged some

astronomers to look at possible alternatives (and, specifically, to one particular alternative) to dark matter, which will be explored in Chapter 6.

To most astronomers, however, the evidence for dark matter seems compelling and overwhelming. The rotation curves of spiral galaxies, the dynamics of galaxy clusters, the x-ray emission from hot clouds of intracluster gas, and gravitational lensing provide the most convincing evidence. Also, the fact that each of these approaches leads to a similar disparity, of around ten to one, between the total mass contained in these systems and the total amount of luminous matter (stars, gas clouds, x-ray emitting gas) that they contain, provides a complementary and self-consistent body of evidence that fits the dark matter paradigm. This evidence is corroborated by measurements of the motions of satellite galaxies and globular clusters, the presence of hot x-ray emitting gas around some elliptical galaxies, recent measurements of weak gravitational lensing, and the possible detection of 'dark galaxies'.

If dark matter dominates the universe, what is its nature? Is it ordinary matter that happens not to be shining, is it something altogether different, or a mixture of the two? As we shall see, observational astronomers, cosmologists, theoretical physicists and experimental particle physicists alike have devoted, and are continuing to devote, immense amounts of time and effort to the task of trying to identify the nature of the mysterious, yet dominant, dark stuff that renders luminous matter little more than a thin icing on a dark and ponderous cosmic cake.

CHAPTER FOUR

The Rise and Fall of MACHOs

We know that dark matter exists because of its gravitational influence on the luminous stuff. But what is it? What is the nature of this mysterious dark stuff which appears to outweigh by far the visible components of galaxies and clusters?

Could it simply be ordinary baryonic matter which happens not to be shining? Perhaps the haloes within which the visible components of galaxics are embedded contain various kinds of objects which emit so little light, or radiation of any other kind, that they cannot be detected by current generations of instruments. For example, if the galactic halo were composed of vast numbers of tiny cold isolated bodies such as asteroids, football-sized lumps of rock or out of date editions of *The Hitchhiker's Guide to the Galaxy*, objects of this kind would be exceedingly difficult to detect. But, there is no plausible mechanism that could have turned most of the primordial baryons into objects of this kind. After all, the heavy elements that are needed to make rocky bodies (or books) were forged in the cores of stars and in supernovae, and make up only a very small fraction of the total mass of ordinary matter in the universe. Unless our entire understanding of the origin and relative proportions of the chemical elements and the way in which rocky bodies are created is completely wrong, footballs, asteroids and old editions of books do not seem to be strong dark matter candidates.

More plausible candidates include extremely dim, low-luminosity objects such as old white dwarfs, dim red dwarfs, neutron stars, black holes, brown dwarfs ('failed stars') and planetary-mass bodies. White dwarfs, neutron stars

and black holes represent different end points for the evolution of stars (see Chapter 1). Only a very tiny fraction of the stars that exist in the Galaxy today are massive enough to collapse into black holes at the end of their lives. However, many astronomers believe that the earliest generation of stars, which ignited around the time when galaxies themselves were forming (so-called Population III stars), included a large proportion of short-lived extremely massive stars, many of which may have become black holes at the end of their brief but spectacular lives, black holes which still could be lurking today within our galactic halo. There could even be a population of Very Massive Objects (VMOs) – black holes with masses in excess of a thousand solar masses that could have formed at an extremely early stage in the assembly of galaxies like our own. Another, more speculative, possibility is that the galactic halo includes a population of primordial, or 'mini', black holes – microscopic black holes which, as Stephen Hawking first suggested, may have been created in the extreme conditions that pervaded during the first microscopic fraction of a second after the Big Bang, around the time when quarks were combining to form baryons and mesons. However, although their existence is a theoretical possibility, none has as yet been detected.

Red dwarfs are cool red main sequence stars which, like the Sun, generate energy by means of hydrogen-burning fusion reactions that take place in their cores, but which are tens or hundreds of thousands of times less luminous than the Sun. Dim they may be, but stars of this kind – with masses in the region of a tenth of a

The object labelled 'B' is a brown dwarf which is about a hundred times fainter than the brighter star, A, to which it is a companion.

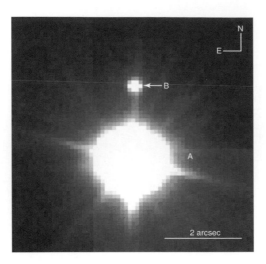

solar mass – are the most abundant and long-lived type of star in our Galaxy. Brown dwarfs are cool dim bodies that are created, like stars, when clouds of gas collapse under the action of gravity but which, because their masses are too low (less than 8 percent of a solar mass), never become hot enough to trigger the nuclear fusion reactions that power normal stars. With surface temperatures typically in the region of a thousand degrees, these 'failed stars' shine dimly at infrared wavelengths by radiating away their reserves of internal heat – heat that was generated by the collapse of the gas cloud, or cloud fragment, from which they formed. If the mass of a collapsed cloud fragment is less than about thirteen times the mass of the planet Jupiter (0.013 solar masses) its nature is more akin to a giant planet, such as Jupiter, than a (failed) star; objects of this kind are referred to as 'planets' or 'Jupiters' (although the use of the word 'planet' to describe a freely floating object in interstellar space, which did not form as part of a star's planetary system, is controversial).

Despite their exceedingly low luminosities, significant numbers of brown dwarfs have been detected in recent years. In the Pleiades star cluster, for example, observations suggest there are as many brown dwarfs as ordinary stars. Indeed, the so-called mass function of stars – a graph of the numbers of stars of different masses – shows that the lower the mass, the more abundant the star. Low-luminosity red dwarfs are the most abundant species

of star, and, if the trend continues, brown dwarfs – which are even less massive – may be more abundant still. Furthermore, a recent infrared survey of the star-forming region in the Orion nebula, revealed, in addition to about a hundred brown dwarfs, about a dozen 'freely-floating planets' – bodies with masses in the region of 8–13 Jupiter masses. These specific examples hint that our Galaxy may well contain at least as many brown dwarfs and 'Jupiters' as stars.

Dark bodies in the galactic halo, such as sub-luminous stars and 'Jupiters', have come to be known as MACHOs – an acronym for MAssive Compact Halo Objects (or, as some prefer, Massive Astrophysical Compact Halo Objects). Since the early 1990s, several different research teams have been searching for MACHOs by looking for the telltale signature of a phenomenon called gravitational microlensing.

As Einstein showed, if a massive body lies between a distant star and an observer, its gravitational field can act like a lens to focus rays of light from that star. If the alignment were absolutely perfect, we would see the image of a star spread out into a ring, known as an Einstein ring, the radius of which depends on the mass of the lensing object. In practice, if the lensing object has a mass comparable to or less than the mass of the Sun, and is located far away in the galactic halo, the resulting Einstein ring is far too small to be resolved by any existing telescope. However, because the gravitational lens pulls rays of light together it concentrates the star's light, causing it to appear brighter. As a lensing object tracks across in front of a background star, it causes the star's apparent brightness to rise to a peak and then, as the 'lens' moves away, drop back down to its original brightness. The increase in brightness depends on the distances of the star and the lens and on the minimum angular separation between the background star and dark lensing object. The phenomenon is called microlensing if it is caused by a foreground object with a relatively low mass – for example, a very faint star, brown dwarf, or planet.

During an event of this kind, the apparent brightness of a star can rise to many times its normal value – more than 40 times, if the alignment is sufficiently close. Its duration depends on the Einstein radius of the lensing object and the speed at which it moves across the sky. For a background star in the Large Magellanic Cloud (some 170,000 light-years away) and a lensing object in the halo of our Galaxy, the duration of an event would be about 10–20 days for a MACHO of 0.01 solar masses, about 200 days for a solar mass object, and about 500 days for a 5 solar mass black hole. An event of this kind would have a highly symmetric light curve (a graph of how brightness changes with time) which, because gravity deflects all wavelengths of light by the same amount, would look the same when measured at two or more different wavelengths (in contrast, a variable star, which changes in temperature as its brightness varies, is likely to brighten by different amounts at different wavelengths). Because microlensing events have such distinctive signatures, they ought to provide a clear and effective way of identifying MACHOs in the galactic halo.

In principle, astronomers can search for MACHO microlensing events by looking along different lines of sight through the halo towards distant background stars. In practice, such searches are extremely difficult to undertake. Brightening due to microlensing can only

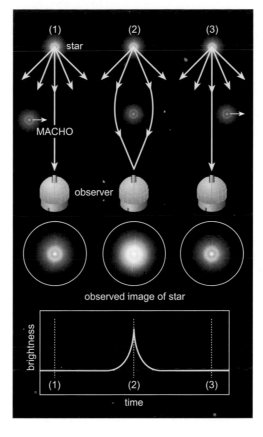

When a MACHO passes in front of a distant star, the focussing effect of the MACHO's gravity causes the star to brighten and fade in a very distinctive way.

be detected if the angle between the moving foreground lens and the background star becomes less than one milliarcsecond (one thousandth of a second), an angle roughly equivalent to the apparent size, as viewed from

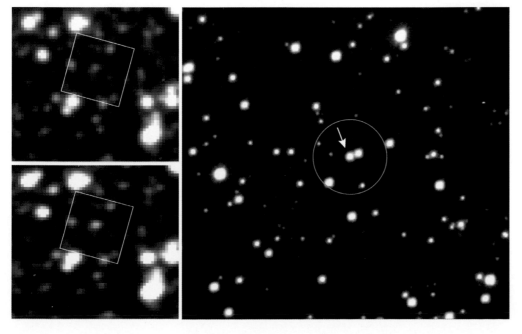

(Left) Two images obtained with a ground-based telescope show the brightening of a star due to gravitational microlensing. (Right) This Hubble Telescope image, taken three years later, resolves the star (arrowed) that was lensed. The lensing object is believed to have been a 5-6 solar mass black hole.

the Earth, of an astronaut standing on the surface of the Moon. The alignment between the source, the lens and the observer has to be so precise that, in order to have any realistic chance of seeing even a few events within a few years of observing time, many millions of stars have to be continuously monitored. To add to the complications, microlensing events caused by genuine MACHOs have to be separated out from a background of similar events caused by ordinary stars that happen to lie along the selected line of sight. Star-on-star lensing – the lensing of a background star by an ordinary foreground star – is (rather confusingly, perhaps) known as 'self-lensing'.

Furthermore, great care is needed to distinguish microlensing events from the brightness variations of variable stars, which are likely to

outnumber lensing events by a factor of at least a hundred thousand. Although microlensing light curves have very distinctive characteristics, supernovae need to be carefully filtered out, and there are many kinds of variable star that are poorly understood or studied and which, potentially, could contaminate the results. The data has to be painstakingly sifted and filtered to weed out and reject imposters.

To probe the galactic halo effectively, the most obvious and convenient sources of large numbers of distant (resolvable) background stars are the Large Magellanic Cloud (LMC) and Small Magellanic Cloud (SMC), nearby satellite systems of the Milky Way Galaxy which lie in the southern hemisphere of the sky. Large-scale monitoring of stars in the LMC got under way in the early 1990s. The two longest-

The central bulge of the Milky Way Galaxy, which lies in the direction of the constellation of Sagittarius, is partly obscured by intervening clouds of dust.

running searches are the joint, Australia-US-UK MACHO project, which utilised the 1.3-metre (50-inch) 'Great Melbourne Telescope' at Mount Stromlo, Australia, and the French EROS (Expérience pour la Recherche d'Objets Sombres) project which used telescopes based at the European Southern Observatory high in the Andes mountain range, in Chile. Another long-running programme – the Polish-American OGLE (Optical Gravitational Lensing Experiment) project, based at Las Campanas Observatory, also in the Andes – has concentrated primarily on looking for microlensing in the direction of the galactic bulge. The first candidate events were detected in 1993.

In 1997, based on data from the first two years of survey operations, the MACHO team published results that appeared to indicate that about 50 percent of the halo consists of MACHOs with masses in the region of half the mass of the Sun, and which did not exclude the possibility that all of the halo mass could be contributed by these objects. The EROS group, which had detected fewer events, contradicted that claim; their results indicated that half-solar mass MACHOs could not make up so great a proportion of the halo. But both groups agreed on one key point: because neither group had detected any events that lasted for less than a month, no more than about 20 percent of the total mass of the halo could be provided by objects ranging in mass from around 10^{-7} solar masses (the mass of the Moon) to around a tenth of a solar mass. This eliminated the possibility that most of the mass of the galactic halo could be contained in brown dwarfs or planetary-mass objects.

Between 1992 and 2000, the MACHO programme monitored a total of 11.9 million stars in regions of the sky towards the galactic bulge and the two Magellanic Clouds. Most of their observed events took place in the direction of the galactic bulge where stars are most densely concentrated and self-lensing is most common. However, by the beginning of the year 2000, on the basis of 5.7 years of data (out of their seven and a half year accumulated total), the MACHO team had identified 13–17 well-defined events along lines of sight through the halo in the direction of the LMC. According to the MACHO team the number of stellar lensing events that would have been expected statistically, based on the known population of stars along the line of sight towards the LMC, was only in the region of 2–4. The durations of the observed events, which ranged from 34 to 230 days, pointed towards MACHO masses from about 0.2 to 0.9 solar masses, with a most probable value of about 0.6 solar masses. Masses in this range are consistent with white dwarfs rather than brown dwarfs or 'planets'. On the basis of these results, the MACHO team concluded that MACHOs contribute about 20 percent of the total mass of the dark galactic halo out to a distance of around 50 kiloparsecs (the distance of the Magellanic Clouds).

By that time, after monitoring the Magellanic Clouds for eight years, the EROS team had only found what they described as a 'meagre crop' of three microlensing candidates towards the LMC and one towards the SMC, despite having monitored a larger number of stars than the MACHO team had done (although each star was monitored less frequently than the MACHO team's stars). Nevertheless, these results were sufficient, in their view, to exclude the possibility that the halo was composed entirely of MACHOs with masses anywhere between about four solar masses and the mass of the Moon. By 2003, the EROS-2 survey, which had started in 1996, had yielded four long-duration candidates towards the SMC, from which they ruled out, with 95 percent confidence, the possibility that MACHOs in the range 2×10^{-7} solar masses to one solar mass could make up more than 25 percent of the mass of the halo (or at least, of the Standard Model of the galactic halo which is assumed by theoreticians). Subsequently, in 2006, the EROS-2 team concluded that MACHOs with masses of between 10^{-7} and five solar masses make up less than seven percent of the overall mass of the galactic halo.

The MACHO results (which implied that about 20 percent of the galactic halo is made up of MACHOs with masses of about half that of the Sun) were seriously challenged in 2004 by Cambridge astrophysicists, Vasily

(Left) This microlensing event in the direction of the galactic centre was observed by the MACHO Project. The light-curves (top and centre) show the changing brightness of the lensed star at blue and red wavelengths. The bottom diagram, which plots the difference between the two curves, shows that they are both virtually identical.

(Right) This image of densely packed stars in the direction of the galactic centre shows, at the centre of the 'crosswires', the location of 'MACHO bulge event 1', the microlensing event shown in the adjacent graphs.

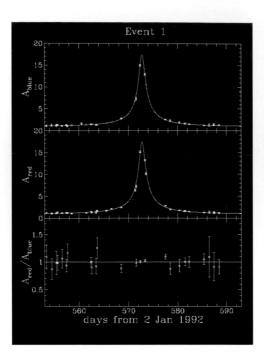

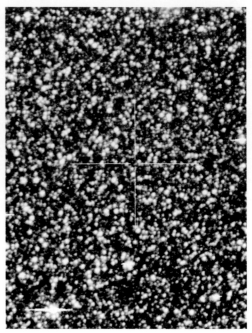

Belokurov and N Wyn Evans and Oxford astrophysicist, Yann Le Du. They re-evaluated a sample of 22,000 MACHO light curves (out of the 11.9 million total) and concluded that the MACHO team's published results were seriously contaminated by non-microlensing events. They asserted that only seven of the original 17 events identified by the MACHO team were genuine microlensing events, and suggested that this smaller number of events could be accounted for by the already known populations of stars in the outer regions of our Galaxy and in the LMC itself, in which case there was no need for any contribution from MACHOs at all. Furthermore, they pointed out that the EROS team had discovered that one of the MACHO team's best candidate objects ('Event 23') had brightened again seven years later and could not, therefore, have been a genuine microlensing event at all (the chance of the same background star being lensed twice in seven years is vanishingly small). At a dark matter conference held in September 2004, Evans and Belokurov went so far as to announce in a paper entitled 'RIP The MACHO Era (1974–2004)' that 'The MACHO era is over! The dark matter in the halo of the Milky Way is not in the form of massive, compact halo objects.'[1]

These claims drew swift rejoinders from members of the MACHO consortium, who hotly disputed Evans and Belokurov's re-analysis and produced higher-quality data on some of the original MACHO events that supported the microlensing interpretation. They did accept, however, that Event 23 was not a genuine microlensing event and, as a result, revised their estimate of the MACHO mass fraction in the galactic halo downwards by 8 percent (from around 20 percent of the total to around 18 percent of the total) – a modest change. But they also conceded that a bigger worry was that there could be more such events in the MACHO sample.

One puzzling feature of the MACHO and EROS results is that they reveal substantially more lensing events in the direction of the LMC than in the direction of the SMC. If the Milky Way's dark halo were spherical (the simplest assumption), we would expect to see marginally more events along the line of sight to the more distant SMC than towards the LMC; yet the observations indicate the opposite. Furthermore, the average duration of the events seen in the LMC by the EROS team is much longer than the average duration of the LMC events that were detected by the MACHO team. Both of these conundrums argue against the hypothesis that all of the events observed towards the LMC and SMC are caused by a common population of MACHOs in the halo of the Milky Way. Instead, the extra events in the direction of the LMC may be due to lensing caused by an as yet undiscovered extended distribution of stars around the LMC itself. Whether the presently available results indicate the presence of MACHOs in the halo of the Milky Way or self-lensing by stars in the Magellanic Clouds is a riddle which, for the moment, remains unresolved.

One of the difficulties associated with trying to detect MACHOs in the halo of the Milky Way Galaxy is the classic 'can't see the wood for the trees' problem – we are inside the system we are trying to explore, and cannot view it as a whole from the outside. To try to resolve the question of what fraction (if any) of the mass of a typical galaxy's halo is contributed by MACHOs, astronomers have started to turn their attention to other galaxies, in particular, the Andromeda galaxy (M31), the nearest large spiral (larger, indeed, than our own), which is tilted almost edge-on towards us. Although lines of sight from the Earth to the Andromeda galaxy probe the halo of our own galaxy and the halo of M31 itself, we nevertheless would expect to see substantially more lensing events along a line that extends over the top of that galaxy's nuclear bulge right through to the far side of its halo, than along a shorter line of sight that leads into its Earth-facing (front) side. Because the Andromeda galaxy is more than ten times further away than the Magellanic Clouds, most of its individual com-ponent stars cannot be resolved (seen as sepa-rate points of light), and each pixel in a CCD image of the galaxy incorporates the combined light of a great many stars. Although this makes it extremely difficult to extract the elusive sig-nature of microlensing, the challenge has been taken up by two ongoing projects – MEGA (Microlensing Exploration of the Galaxy and Andromeda) and POINT-AGAPE (Andromeda Galaxy Amplified Pixel Experiment).

The results so far are somewhat contradic-tory. From three years of data acquired using a wide-field camera on the 2.5-metre Isaac

Newton Telescope at La Palma in the Canary Isles, the POINT-AGAPE team identified six short-period lensing events along lines of sight towards the Andromeda galaxy. Even after excluding one of them, which probably involved a star that lies within M32 (one of the Andromeda galaxy's elliptical satellites, which happens to lie along the chosen line of sight), the team concluded, in the summer of 2005, that they were observing far more events than could be explained by self-lensing (star-on-star lensing) caused by known populations of stars in the Andromeda galaxy or in the Milky Way's halo. Their prediction for self-lensing was less than one event, whereas they actually observed five. They were confident, therefore, that they were detecting genuine MACHO events, and concluded that if the masses of these MACHOs lay between 0.5 and 1 solar masses, then MACHOs must contribute at least 20 percent of the halo mass along that particular direction – a result that matched quite well with the conclusion that the MACHO team had arrived at some five years previously.

The MEGA collaboration has been looking for MACHOs in the Andromeda galaxy's halo using four different telescopes, including the Isaac Newton Telescope (INT) on La Palma and the 4-metre Mayall Telescope at Kitt Peak, Arizona. In the summer of 2005 they too published the outcome of four years of observations made specifically with the INT. Although they identified a total of 14 candidate microlensing events, they contended that most, if not all of them, could have been caused, not by MACHOs, but by self-lensing. As a result, they concluded that the fraction of Andromeda's halo that is made up of MACHOs was most probably somewhere between zero and 10 percent and were confident (at the 95 percent confidence level, if you are of a statistical bent) that their data ruled out the possibility that MACHOs could make up more than 30 percent of the mass of the halo. Their bottom line was that 'we find no compelling evidence for the presence of MACHOs in the halo of M31'.

So, for the moment, the view towards the Andromeda galaxy presents a rather conflicting

This artist's impression shows the GAIA spacecraft, which is currently under development, mapping the stars of the Milky Way. Its mission will be to produce a precise three-dimensional map of more than a billion stars in our Galaxy and beyond.

and confused picture. On the one hand, the POINT-AGAPE team reckon their evidence points to at least 20 percent of the halo mass in the direction of M31 being in the form of MACHOs, whereas, on the other, the MEGA team claims that MACHOs contribute less that 30 percent, and most probably between zero and 10 percent of the M31 halo. The picture may sharpen a little when further results from MEGA and POINT-AGAPE roll in, or when sufficient results have been accumulated by SuperMacho, the successor to the MACHO project, which is using a high-resolution, wide-angle camera (the MOSAIC Imager) mounted on the 'Blanco' 4-metre telescope, high in the Chilean Andes, to monitor tens of millions of stars in the LMC.

The ultimate test may come eventually from space missions such as NASA's Space Interferometry Mission (SIM), or the European Space Agency's GAIA mission (Global Astrometric Interferometer for Astrophysics), each of which is scheduled for launching in 2011. Designed to measure the positions and parallaxes of stars to an accuracy of 10 micro-arcseconds or better (an angle of 10 micro-arcsec – 10 millionths of a second of angular measurement – is equivalent to the width of a human hair seen from a distance of 1000 kilo-metres), these missions should be capable of measuring the distances of stars to an accuracy of 10 percent or better anywhere within the

Galaxy. Each mission has the potential to measure the distance of at least some of the microlensing objects. If that can be achieved, the results would at least show conclusively whether the extra microlensing along lines of sight towards the LMC is due to our Galaxy's halo or to lenses in the LMC itself, and likewise would show whether lensing events along the line of sight to the Andromeda galaxy were, or were not, due at least in part to objects in our own galaxy's halo.

Although the current observational picture remains a shade blurry, it does look as if compact baryonic MACHOs with masses ranging between 10^{-7} solar masses (the mass of the Moon) and around 30 solar masses have been ruled out of contention as the principal contributors to the overall mass of the Galaxy's dark halo. Regardless of whether MACHOs contribute about 20 percent of the mass of the halo, almost nothing at all, or some figure between those values, all of the protagonists agree on a key central point: MACHOs alone cannot account for the dark matter halo of our Galaxy, still less the dark matter problem in the universe at large. The answer must lie with something else.

It's Matter - But Not as we Know it

The evidence is very clear: clumps of baryons in the form of MACHOs cannot account for more than about 20 percent of the mass contained in the dark haloes of galaxies. While it is still possible that baryons in some other guise, such as black holes, or cool molecules (or even solid flakes) of hydrogen, could make some kind of significant contribution to the overall masses of galaxy haloes, there are powerful theoretical arguments and observational evidence which show beyond all reasonable doubt that the universe does not and cannot contain enough baryonic matter (luminous or dark) to account for more than a modest fraction of the total amount of dark matter that the universe contains.

The key theoretical argument hinges around Big Bang nucleosynthesis – the process whereby some of the primordial hydrogen was converted into elements such as helium and lithium during the first few minutes of the history of the universe. As we saw in Chapter 2, by the end of this flurry of activity, baryonic matter consisted almost entirely of hydrogen and helium nuclei (the mix contained 11 hydrogen nuclei to each helium nucleus), together with a tiny proportion of the other fusion products – deuterium (heavy hydrogen), helium-3 (the lighter isotope of helium), and lithium. Because the exact proportion of each chemical element and isotope that emerged from this process depended in a sensitive way on the overall baryon density at the time when these reactions were taking place, astronomers can use measurements of the observed relative abundances of these elements and isotopes to calculate the mean density of baryonic matter in the universe today. It turns out that in order to account for the observed relative proportions of these elements, the total amount of baryonic matter in the universe cannot be more than about 4–5 percent of the critical density.

This argument has been reinforced by other kinds of measurements. In particular, precise measurements of warmer and cooler patches in the cosmic microwave background radiation, which mirror the way in which matter was gathered into marginally denser and more rarefied clumps when the universe was about 380,000 years old, have shown that baryons contribute just 4.4 percent of the critical density (see Chapter 8). Detailed studies of the way in which matter, on the large scale, is aggregated into galaxies, clusters, filaments and sheets, and the way in which these large-scale structures have grown with the passage of time, have produced closely similar results.

As we have seen, many of the largest clusters of galaxies are permeated by extensive clouds of exceedingly hot ionised gas (intracluster gas) which shines at x-ray wavelengths. As we saw in Chapter 3, astronomers can use measurements of the size, x-ray brightness and temperature of these clouds to measure the overall mass of the gas itself and the total mass of the cluster (including dark matter). By adding the mass of the hot intracluster gas to the optically luminous mass of its constituent galaxies, they can calculate the total amount of baryonic matter in the cluster. The results show that the total mass of a typical large cluster is about six or seven times greater than its baryonic mass. Since many astronomers reckon that the matter content of the largest galaxy clusters represents a reasonably fair sample of the matter content of the universe as a whole, it is therefore reasonable to assume that the observed ratio of baryonic mass to total mass in these clusters will be similar to the ratio of baryonic-to-total mass in the universe at large.

If that is the case, then the average matter density in the universe can be obtained by multiplying that ratio by the mean baryon density derived from Big Bang nucleosynthesis (which is about 4.4 percent of the critical density). After making allowance for any baryons which were expelled from clusters while they were forming, the data from x-ray clusters implies that the mean density of matter (both luminous and dark) is about 25 percent of the critical density; the ratio of the total matter density to the critical density is $\Omega_M \approx 0.25$. Studies of the cosmic microwave background radiation and of the large-scale distribution of matter in the universe point to very similar values (see Chapter 8); for example, results obtained during the first three years of operation of the Wilkinson Microwave Anisotropy Probe (WMAP) and published in 2006, give an overall matter density equal to 26 percent of the critical density ($\Omega_M = 0.26$), which is about six times greater than the baryonic density ($\Omega_{baryons} = \Omega_b = 0.04$).

Collectively, these results show that baryonic matter comprises only about 15 percent of the total amount of matter in the universe. The remaining 85 percent or so is non-baryonic – it is not composed of protons and neutrons, and is completely different in nature to the stuff of which tables, chairs, planets, stars and people are made.

What is this mysterious non-baryonic stuff that appears to comprise around 85 percent of the total matter density of the universe? Most probably it consists of vast numbers of exotic elementary particles which, because they are not baryons, do not respond to the strong nuclear force that binds quarks and atomic nuclei together. Like ordinary matter particles, they have mass and are acted on by

the exceedingly feeble force of gravity, but otherwise are acted on only by the weak (or the electroweak) force. Consequently, they interact directly with particles of ordinary matter and photons so exceedingly rarely that they are almost impossible to detect. Candidate particles include neutrinos and WIMPs (an acronym for Weakly-Interacting Massive Particles). Neutrinos are known to exist. WIMPs, by contrast, have been predicted theoretically, but as yet none has been definitely detected.

The little neutral one

Neutrinos are electrically neutral elementary particles which are acted on only by the weak nuclear force and gravity. Austrian physicist, Wolfgang Pauli, proposed their existence in 1931 in order to resolve a budgetary problem in particle physics that came to light during the late 1920s, when physicists were studying the radioactive decay of certain kinds of atoms through a process known as beta decay. In beta decay an electrically neutral neutron changes into a positively-charged proton through the emission of a negatively-charged electron (electrons were known as beta particles – hence the name 'beta decay'). Experimenters were puzzled to find that the emerging electron and proton seemed to have less energy than expected, as if energy had mysteriously vanished. In order to resolve this conundrum, Pauli postulated the existence of another, unseen, particle which would carry away some of the energy and thereby 'balance the books'. The proposed particle had a rather bizarre set of properties. As originally conceived, it had zero electrical charge and zero rest-mass (a stationary neutrino would weigh nothing at all), but nevertheless possessed a finite amount of a quantity called spin angular momentum (it could be visualised as spinning round an axis like a tiny ball – see Chapter 1). Like a photon, it would travel at the speed of light, and while moving, carry with it a finite amount of energy. Some two years later, Italian physicist Enrico Fermi gave the postulated particle the name 'neutrino', Italian for 'little neutral one'.

Neutrinos interact so feebly with matter that they are exceedingly hard to detect.

Consequently, a quarter of a century elapsed before the existence of these elusive particles was confirmed experimentally by American physicists, Clyde Cowan and Fred Reines. Cowan and Reines set up an experiment at the Savannah River nuclear reactor in the USA – appropriately entitled, given the elusiveness of their quarry, 'Project Poltergeist' – to detect inverse beta decay, the process whereby a proton captures an antineutrino (like other particles, neutrinos have their antiparticles) to produce a neutron and a positron (an anti-electron). The nuclear reactor generated such a large outpouring of neutrinos that very occasionally a neutrino ought to have interacted with a suitable target atom (Reines and Cowan used tanks of cadmium chloride as targets). They had calculated that the key signature of an event of this kind would be the release of two energetic gamma-ray photons, separated by a time interval of 5 microseconds and, in the summer of 1956, this is exactly what they found. The existence of the neutrino had been confirmed.

Neutrino families and the Standard Model

The neutrinos involved in beta decay were associated with electrons. But neutrinos were subsequently shown to be associated also with heavier analogues of the electron – muons (which are about 400 times more massive than electrons), and tau particles (which are about 3500 times heavier than the electron). We now know that there are three types, or 'flavours', of neutrino – the electron-neutrino, the muon-neutrino and the tau-neutrino.

Neutrinos form an integral part of the Standard Model of particle physics, which was developed between the 1960s and 1980s. The Standard Model provides a simple descrip-tion of the elementary particles and forces in terms of matter particles (collectively called fermions) and force-carrying particles, called bosons, which 'mediate', or transmit, the forces of nature between matter particles (see Chapter 1). The Standard Model particles con-sist of six quarks (point-like particles which are acted on by the strong nuclear force) and six

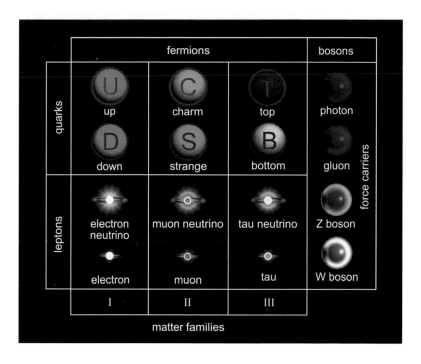

leptons (particles that do not experience the strong nuclear force). Matter particles (fermi-ons) are arranged in three families – each fam-ily containing two quarks and two leptons, one of which is a neutrino. The first family consists of the 'up' and 'down' quark (of which protons and neutrons, and all the familiar baryonic mat-ter of everyday experience, is composed) and the electron, together with a neutrino, called the electron-neutrino (the electron-family neu-trino). The second family consists of the more exotic 'strange' and 'charm' quarks, the muon and the muon-neutrino; the third family con-sists of the 'top' and 'bottom' quark, the tau, and the tau-neutrino. Each of these particles has an equivalent antiparticle, with opposite charge and spin.

In addition, there are four force-carrying bosons: the photon, which transmits the electromagnetic force, the gluon, which con-veys the strong nuclear interaction that binds quarks together inside protons and neutrons, and the 'W' and 'Z' bosons (W bosons have positive or negative charge, whereas the Z boson is neutral), which convey the weak nuclear interaction that governs processes such as radioactive decay. Whereas the photon and gluon are massless particles, the W and Z bosons weigh in at around 80–90 times the

The Standard Model arranges matter particles (collectively called fermions) into three families, each consisting of two quarks and two leptons (plus their antiparticles), and also includes four force-carrying particles (bosons).

mass of the proton. In order to give the W and Z bosons finite masses, the Standard Model requires an additional boson, called the Higgs boson, whose existence was proposed in 1964 by Edinburgh physicist Peter Higgs. Without it, the Standard Model implies that all particles would be massless (like the photon and gluon). The hypothesised Higgs particle has zero spin and is believed to weigh somewhere between 100 and 300 times the mass of the proton. As yet, the Higgs boson has not been detected experimentally.

The Standard Model has been very successful in many respects. For example, it explains how elementary particles interact with each other and, indeed, it predicted the existence of many of its family of particles before they were detected. However, particle physicists recognise that, despite its considerable successes, the Standard Model is not a complete or final theory. It does not resolve a number of current problems in particle physics (for example, it implies that particles and antiparticles ought to exist in equal numbers, whereas the observed universe consists almost entirely of matter rather than antimatter) and it does not incorporate gravity.

The ultimate goal is to develop a 'Theory of Everything' which embraces gravity and unifies all the forces. Unification implies that at progressively higher energies, the strengths of the forces would begin to converge, and at sufficiently high particle energies (such as would have pertained in the very earliest instants of the Big Bang), all four forces would merge into a single unified 'superforce'. Running forward in time from the Big Bang, the four forces are believed to have separated out as the universe expanded and cooled and as the energies of particles declined. Gravity peeled off first. Then the electroweak force (a unified combination of the weak and electromagnetic forces) separated out from the strong force. Finally, around 10^{-10} seconds after the beginning of time, when temperatures had dropped to around 10^{15} K, the weak and electromagnetic forces revealed their separate identities. The validity of electroweak theory was demonstrated in 1983 by the discovery of the W and Z bosons

(the neutral Z boson is a key signature of the electroweak force) in detectors associated with a powerful particle accelerator called the Super Proton Synchrotron at the CERN (Conseil Européen pour la Recherche Nucléaire) laboratory on the outskirts of Geneva. But whereas experimenters have been able to achieve energies high enough to attain electroweak unification, other unifications will only occur at energies far in excess of anything attainable by man-made particle accelerators.

Just a little mass

Neutrinos would have been created in great abundance in the early instants of the Big Bang, and would have ceased to interact with ordinary matter and radiation (and therefore to have been created and destroyed) by about one second after the beginning of time. Thereafter this population of primordial neutrinos would have spread out through the expanding volume of space, hardly ever interacting with particles of matter or with photons. Theory suggests that in the universe today there ought to be about 100 million to 200 million neutrinos in every cubic metre of space – a cosmic neutrino background analogous to the microwave background. So long as neutrinos have zero mass, they make hardly any impact on the overall mass-energy density of the universe, but if instead neutrinos have tiny, but finite masses, then there are so many of them around that they need only have a few hundred thousandths of the mass of an electron (which itself has just over one two-thousandth of the mass of a proton) to account completely for the known amount of dark matter. If the neutrino mass were in the region of one ten thousandth of the electron mass, neutrinos would provide enough matter density to 'close' the universe – to ensure that it would eventually collapse.

According to the Standard Model, all three flavours of neutrino have zero mass. But because the Standard Model has its limitations, and cannot account for all aspects of the universe in which we live, particle physicists have for some time been willing to entertain the possibility that neutrinos might have

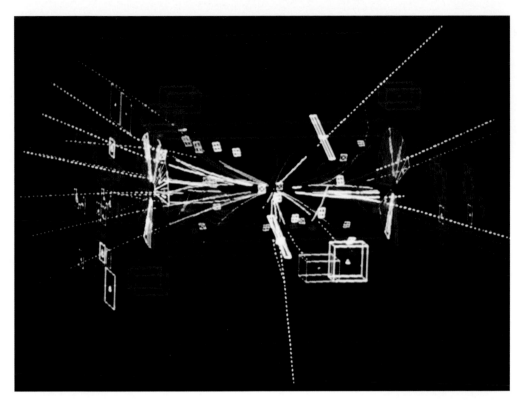

Decay of a Z-boson – the uncharged carrier of the weak force – into an electron and a positron seen for the first time on 30 April 1983 at the UA1 detectors at CERN. The two white tracks extending from the centre towards the upper right, and almost directly downwards are the tracks of the high-energy electron and positron that carry the mass that had once been the short-lived Z particle.

microscopic but finite masses. Observational and experimental hints that this might be so began to emerge in the 1960s when astronomers came up against the so-called solar neutrino problem. The nuclear reactions which convert hydrogen to helium in the core of the Sun release neutrinos, most of which – travelling at, or very close to, the speed of light – shoot unimpeded through the body of the Sun and escape to space. A small fraction of the solar neutrinos encounter the Earth and, again, all but a tiny fraction of them pass straight through our planet and out the other side without noticing. Indeed, as you read this page, about 10 trillion neutrinos are passing through your body every second, just as if you were not there. Because the Earth itself is essentially transparent to neutrinos, as many solar neutrinos will pass through your body at midnight (when the Sun is on the far side of the Earth) as at midday, when the Sun is high in the sky.

Detecting neutrinos is a difficult and challenging task. Whereas an x-ray can be stopped with a few centimetres of lead, a block of lead light-years thick would be needed to stop the average individual neutrino! Nevertheless, if a detector contains a large enough number of suitable target atoms it should be able to capture a microscopic fraction of the prodigious flow of solar neutrinos. The likelihood of an interaction between a neutrino and an atomic nucleus is described by a quantity called the cross-section which, in essence, is the effective target area presented by a nucleus to an incoming neutrino. For most atomic nuclei, the target area is so tiny that for a column of rock stretching from one side of the Earth to the other, an incoming neutrino has less than a one in a trillion chance of being stopped by a collision.

The first successful experiment of this kind, which was set up in 1965 by Professor Ray Davis, of Brookhaven National Laboratory, used a heavy isotope of chlorine (chlorine-37) to capture neutrinos. The cross-section of ^{37}Cl, for the neutrinos to which it is sensitive, is only about 10^{-42} centimetres squared, so that for a capture to occur, a neutrino in effect has to hit a 'bulls eye' with a diameter a billion times smaller than the nucleus of an atom. The Davis detector consisted essentially of a tank filled

This image of the
Brookhaven solar neutrino
experiment, shows the
400,000 litre tank of
perchloroethylene which was
used to capture neutrinos,
and which was located
1500 metres underground
at the Homestake Mine in
South Dakota. Reproduced
courtesy of Brookhaven
National Laboratory.

with about 400,000 litres of perchlorethylene
(a form of dry-cleaning fluid but consisting
mainly of liquid chlorine), a quantity of liquid
that contains about 2.2×10^{30} chlorine-37
atoms. When a ^{37}Cl nucleus captures a neutrino
it is converted into a radioactive form of argon.
The dissolved argon atoms are then removed
from the liquid and counted by detecting the
electrons that are emitted when they undergo
radioactive decay. Knowing the capture cross-
section of chlorine-37, the flux of neutrinos at

the Earth, and hence the rate at which neutrinos were being produced in the core of the
Sun, could be calculated.

In order to minimise contamination by particles such as cosmic rays (highly energetic subatomic particles and atomic nuclei which pour
down on the Earth from space), which could
mimic the effects of neutrinos, the instrument
had to be heavily shielded. It was placed at a
depth of 1500 metres in the Homestake Gold
Mine in South Dakota so that the overlying

rock would cut out cosmic rays and other particles – only neutrinos could penetrate freely through such depth of rock. Even so, natural radioactivity in the surrounding rock was a major problem and had to be minimised by surrounding the tank by further shielding in the form of a water jacket. The ingenuity involved in setting up such an experiment deep down in a mine was extraordinary – and similar strategies have since been adopted not only by seekers of neutrinos but by experimenters hunting for other kinds of dark matter particles, such as WIMPs, as we shall explore in Chapter 7.

Right from the outset the experiment produced perplexing results. Too few neutrinos were being detected. At first, critics suggested there might be problems with the experimental apparatus, methodology, the extraction of the argon, and so on. One problem is that the chlorine detector was not sensitive to the low-energy neutrinos which are released at the first stage of each and every complete hydrogen fusion reaction in the Sun. Instead it was sensitive mainly to relatively high-energy neutrinos that, according to theory, are released in only about one out of every ten thousand solar fusion reactions. But as years went by, and stringent tests were applied to the chlorine detector, and as other kinds of neutrino detectors using different target atoms (for example, gallium) began to find the same shortfall, the scientific community was forced to take seriously what the experimenters were finding – only about one-third of the expected numbers of solar neutrinos appeared to be reaching the Earth. The shortfall in detected neutrinos came to be known as the solar neutrino problem. The question was: did the results indicate that there was a problem with the theory of nuclear energy generation in the Sun, a problem with the experiments, or a problem with the Standard Model of particle physics?

Gradually, a potential solution began to emerge and gain ground. There are three different types, or flavour, of neutrino. If neutrinos could change from one type to another, and oscillate between the three different types, a batch of one particular flavour would convert into a mixture of flavours. The fusion reaction which powers the Sun produces electron neutrinos. If, while en route from the Sun to the Earth (or, more likely, while travelling out through the body of the Sun), the original batch of electron-neutrinos turned into a mix containing roughly equal proportions of the three different neutrino flavours, then those types of neutrino detector (such as the Davis chlorine detector) which were sensitive only to electron-neutrinos would register only about one-third of the predicted flux of solar neutrinos. If neutrinos have zero mass, as the Standard Model implies, flavour mixing would not occur (if their masses were zero, neutrinos would travel at the speed of light and would not experience the passage of time: therefore they could not oscillate between flavours). For flavour-mixing to occur, neutrinos have to have finite, albeit tiny, masses.

The first key breakthrough in confirming that neutrinos do indeed have mass, and oscillate between different flavours, came in 1998 as a result of work carried out by a team of Japanese and US physicists using a neutrino detector called Super-Kamiokande, which is located about a kilometre underground in the Kamioka Mining and Smelting Company's Mozumi Mine, some 300 kilometres west of Tokyo. The heart of this detector is a 50,000 tonne tank of ultra-pure water within which are suspended more than 13,000 photomultipliers – devices which record the flashes of light that are emitted on those rare occasions when neutrinos interact with water nuclei in the tank.

The reason that light is emitted hinges on the fact that the speed of light in water is lower than the speed of light in a vacuum. If an incoming neutrino collides with an electron in the water tank, and if the neutrino is carrying enough energy, the electron will recoil at a speed which initially is greater than the speed of light in water. Just as a supersonic aircraft, travelling through air faster than a pressure (sound) wave can propagate, causes a build-up of pressure called a shock wave (which causes a sonic boom when it strikes the ground), so a charged particle travelling faster than the

Technicians in an inflatable boat inside Super-Kamiokande water-based neutrino detector are inspecting some of the 13,000 photomultiplier tubes which are used to detect the flashes of light that are emitted when neutrinos interact with water.

speed of light in the medium through which it is moving, creates a wave of photons – a sort of shock wave of visible light which spreads out in a cone along the direction in which the particle is moving. Light created by this mechanism was first detected in 1934 by the Russian physicist Pavel Čerenkov and is known as

Čerenkov radiation. Water Čerenkov detectors such as Super-Kamiokande, detect neutrinos by recording Čerenkov radiation emitted by recoiling electrons and, because the radiation is directional, are able to measure the direction from which each incident neutrino has arrived. The Super-Kamiokande detector is sensitive to

electron-neutrinos and muon-neutrinos, but cannot detect tau-neutrinos.

When cosmic rays collide with the atmosphere, they generate showers of secondary particles that rain down on to the Earth's surface. One of the particles produced in this way is the pion, which then decays into a muon and a muon-neutrino. The muon itself then decays into an electron, another muon-neutrino and an electron-neutrino, so that the net result is to produce two muon-neutrinos and one electron-neutrino. Therefore, ground-based detectors ought to record twice as many muon-neutrinos as electron-neutrinos. But by 1985 experiments had revealed that the actual ratio was less than two to one – it seemed as if either fewer muon-neutrinos, or too many electron-neutrinos, were being produced. This discrepancy became known as the atmospheric neutrino problem.

In 1998, the Super-Kamiokande team showed that the atmospheric neutrino problem could be resolved by neutrino oscillations. Because neutrinos travel virtually unimpeded through the body of the Earth, virtually the same number of neutrinos per second should pass upwards through the detector, having been produced in the atmosphere on the opposite side of the Earth, as downwards through the detector, having been produced in the atmosphere directly overhead. In the absence of neutrino oscillation, the muon-neutrino flux from above should be almost exactly the same as the muon-neutrino flux from below. However, if neutrino oscillation is taking place, the extent to which a beam of neutrinos transforms into a mixture of flavours depends on how far and for how long it has travelled since being created. Downward-moving neutrinos, having traversed a mere 40 kilometres or so, have not travelled far enough for much flavour-mixing to occur, whereas upward-moving neutrinos, having travelled more than 12,000 kilometres through the Earth, have had more opportunity to mix. If a proportion of the upward-moving muon-neutrinos, while travelling through the Earth, transform into tau-neutrinos, which cannot be detected by Super-Kamiokande, the result would be an apparent shortfall in the detected numbers of muon-neutrinos. The Super-Kamiokande results indicated that the measured muon-neutrino deficit was due largely to the proportion of muon-neutrinos that had converted to tau-

When a charged particle ploughs through a tank of water faster than the speed of light in water, it causes blue light, called Čerenkov radiation, to spread out in a cone centred on the track of that particle. The short arrows show the direction in which photons are heading.

(Left) Recorded by the Super-Kamiokande detector, this ring of Čerenkov light was produced by an electron that was hurtling through water faster than light does, as a result of a collision with an incoming electron-neutrino.

(Right) A muon-neutrino has interacted with water in the Super-Kamiokande detector to create a muon, which produced this ring of light as it moved through the water faster than light. The heavier muon-neutrino produces a sharper-edged ring of Čerenkov light than an electron-neutrino makes.

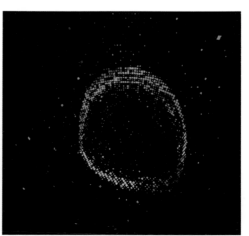

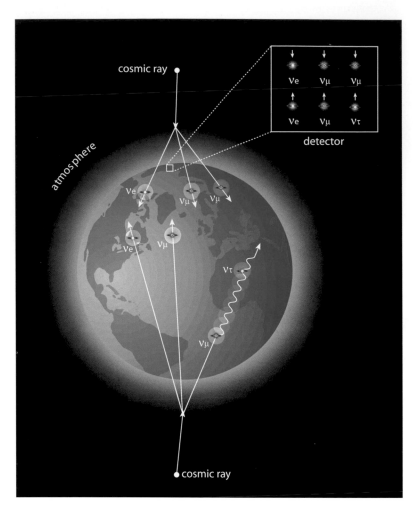

current reactions (whereby a neutrino interacts with a deuterium nucleus to produce a proton, a neutron and a neutrino), and elastic scattering (where in effect a neutrino and an electron bounce off each other). Charged current reactions are sensitive only to electron neutrinos whereas neutral current reactions have equal sensitivity to all three neutrino flavours. Elastic scattering is also sensitive to all three flavours but is considerably less sensitive to muon neutrinos and tau-neutrinos than to electron neutrinos. The 2001 SNO results showed that the detector was registering more elastic scattering events than charged current events, which implied that the total number of solar neutrinos registered by the detector was greater than the number of electron neutrinos alone.

Any lingering doubts about neutrino oscillations were swept away by more sensitive SNO results, published in 2003, which compared the flux of electron neutrinos revealed by charged current events with the total neutrino flux revealed by the neutral current reaction. These measurements confirmed that the total flux of all three neutrino types matches the theoretically predicted flux of neutrinos from the Sun and that electron neutrinos arriving at the Earth make up one-third of that total. These results showed that it is 99.999 percent probable that electron neutrinos mutate into the other flavours en route from the Sun and that neutrinos must, therefore, have finite masses.

Weighing neutrinos

Although Super-Kamiokande, SNO, and other experiments such as KamLAND (the Kamioka Liquid scintillator Antineutrino Detector which studies antineutrinos released by a network of Japanese nuclear reactors) have shown beyond all reasonable doubt that neutrinos do indeed have finite masses, none of them has been able to measure the actual values of their masses. Direct attempts to measure the mass of the neutrino have focused primarily on studying the beta decay of tritium (a heavier version of hydrogen that has one proton and two neutrons in its nucleus) through the release of an electron and an electron-antineutrino, and

Neutrino oscillations solve the atmospheric neutrino problem. When an incoming cosmic ray interacts with the atmosphere it produces one electron-neutrino and two muon-neutrinos. While travelling through the Earth, one of the muon-neutrinos transforms into a tau-neutrino; therefore the detector (inset top right) 'sees' fewer muon-neutrinos than expected.

neutrinos on their way through the Earth. This confirmation of neutrino oscillation showed that neutrinos do indeed have mass.

Strong support for neutrino oscillation came in 2001 from the first set of results released by the Sudbury Neutrino Observatory (SNO) in Ontario, Canada. The target material in this experiment is 1000 tonnes of heavy water, a liquid which is chemically the same as normal water, but with the two hydrogen atoms in each molecule (normally H_2O) replaced by deuterium atoms (like ordinary hydrogen, a deuterium atom has one orbiting electron; however, in addition to a single proton, its nucleus contains a single neutron, which makes it heavier). The heavy water detects solar neutrinos through three different processes: charged current reactions, in which a neutrino interacts with a deuterium nucleus to produce two protons and an electron; neutral on looking for evidence of a process known as neutrinoless double beta decay. Double beta decay is a process whereby certain kinds of nuclei, such as germanium-76, simultaneously convert two neutrons into protons, with the release of two electrons and two antineutrinos. It is possible that the neutrino may in fact be its own antiparticle. If so, the two (anti) neutrinos would annihilate almost immediately after they are produced, and the observed outcome would be the release of two electrons without any accompanying neutrinos. Detection of this phenomenon would indicate that neutrinos are their own antiparticles and would allow their masses to be calculated. So far these approaches have yielded upper limits (maximum possible values) for the mass of the electron neutrino in the region of 2.2 eV/c² (less than a two hundred thousandth of the mass of the electron) in the case of tritium beta decay, and 0.1 to 0.8 eV/c² (with a best value of 0.4 eV/c²) from claimed, but still disputed, detections of neutrinoless double beta decay.

Another route to finding neutrino masses is to seek out signs of their influence in cosmological observations. If neutrinos have finite masses, they will have exerted detectable effects on the way in which the first clumps of matter (primordial density fluctuations) formed and then evolved to generate the patterns of structure that we see in the universe today. The extent of their influence depends on the masses of the neutrinos themselves and on the magnitude of their contribution to the overall density of the universe. Evidence of their influence can be found in subtle modifications of the sizes and strengths of warmer and cooler patches in the cosmic microwave background; in the distribution of galaxies as revealed by large-scale surveys such as the Two-degree Field Galaxy Redshift Survey (2dFGRS) and the Sloan Digital Sky Survey (SDSS); in the distribution of hydrogen gas in the early universe as revealed by the imprint of intervening gas clouds on the spectra of extremely distant quasars (the so-called 'Lyman-alpha forest' – see Chapters 8 and 12); in weak gravitational lensing effects on the observed appearance of galaxies caused by the overall distribution of

matter – baryonic and non-baryonic – along lines of sight between ourselves and distant galaxies; and in the pattern of individual 'peculiar velocities' of galaxies and clusters which is superimposed on the Hubble law of the expansion of the universe

At present, the most stringent limits on the maximum possible mass for the neutrino are provided by cosmological measurements. Detailed analysis of the observed temperature fluctuations in the cosmic microwave background and of the evolution of structure in the universe, as mapped out by large-scale surveys like 2dFGRS and SDSS, have yielded upper limits of less than 1 eV/c² on the sum of

This artist's impression of the Sudbury Neutrino Observatory shows the 12-metre acrylic vessel, which contains the 1000 tonnes of heavy water that acts as the neutrino target, surrounded by a geodesic support structure that holds 9600 photomultiplier tubes. These components, in turn, are immersed in normal water within a 30-metre barrel-shaped cavity.

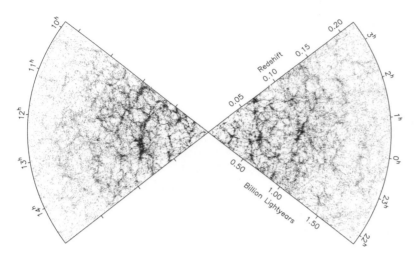

Shown here is the large-scale distribution of galaxies in a cone centred on the Earth and extending out to a distance of nearly 2 billion light-years, as determined by the Two-Degree Field Galaxy Redshift Survey.

the masses of the three neutrino types. The 2dFGRS on its own implies that the sum of the three neutrino masses is less than 2.2 eV/c^2 and that each flavour of neutrino has a mass of less than about 0.7 eV/c^2. By 2004, the combination of WMAP, SDSS and Lyman-alpha forest data had produced a strong upper limit of 0.42 eV/c^2 (less than one millionth of the mass of the electron) for the sum of the three neutrino masses.

In order to contribute enough mass to make up the critical density, the sum of the neutrino masses would have to be about 47 eV/c^2; even to provide 25 percent of the critical density (enough to account for all the non-baryonic dark matter) would require a sum of neutrino masses of around 12 eV/c^2. Taken together, these results imply that neutrinos are unlikely to contribute more than about one percent of the critical density and cannot themselves make a major contribution to the total dark matter budget of the universe. The results seem clear-cut: neutrinos are minor players in the overall mass-energy budget of the universe. But a note of caution needs to be borne in mind; all of these limits are based on the current consensus model of the universe and on the assumption that there are indeed only three neutrino flavours.

Cold dark matter enters the fray

The idea that the universe might be dominated by relic neutrinos (neutrinos created in the Big Bang fireball) had begun to fall out of favour as

early as the 1980s. The reason for their falling from grace hinged on the fact that neutrinos are an example of 'hot' dark matter. When neutrinos were created in the primordial fireball, they would have been moving extremely rapidly (hence the term, 'hot') – at very large fractions of the speed of light – relative to particles of ordinary baryonic matter. Because of their exceedingly high speeds, they would have streamed freely out of any incipient localised concentrations of matter. Also, if hot neutrino dark matter had been the dominant mass component of the universe, the sea of fast-moving neutrinos would have damped out any concentrations of baryonic matter until such time as neutrinos themselves had slowed sufficiently to aggregate together, under the action of gravity, into massive clumps. These clumps would then provide the underlying 'wells' of gravitational influence into which ordinary baryonic matter would 'fall' to create structures such as galaxies and clusters. This 'free streaming' property of hot neutrino dark matter implies that the first structures to be able to hold themselves together under the action of gravity would have been extremely large and massive ones (comparable in mass to galaxy superclusters) that subsequently would have fragmented into clusters and individual galaxies. Structure formation would have proceeded in a 'top down' fashion from the largest to the smallest.

Detailed numerical calculations of the way in which matter would aggregate together under the action of gravity to form structures such as galaxies, clusters, superclusters and voids, carried out using powerful computers, were pioneered in the early 1980s by cosmologists Marc Davis, Simon White, Carlos Frenk and George Efstathiou. Based on the assumption that the universe is dominated by hot dark matter, simulations of this kind generated much stronger patterns of large scale structures in the distribution of matter than are actually seen in the real universe. But if instead the bulk of the dark matter were assumed to be 'cold' – consisting of relatively slow-moving dark matter particles which could clump together much more readily than hot dark matter - computer models produced structures, built from the bottom up (small structures forming first, then aggregating into larger ones), which more closely matched what the large scale features of the universe actually looks like. Furthermore, the balance of observational evidence suggests that, rather that happening in a 'top down' fashion (as the hot dark matter model implied), structure formation proceeded in the opposite direction, with small protogalaxies merging to form progressively larger galaxies that then aggregated together to form still larger entities such as clusters and superclusters. By the late 1980s, arguments such as these had led many cosmologists – as we shall see later – to favour a cold dark matter universe rather than one dominated by neutrinos.

Cold dark matter (CDM) consists of hypothesised elementary particles, produced in the Big Bang, which moved relatively slowly (hence 'cold') relative to neighbouring baryons. Because of this, and the fact that they effectively ceased to interact with ordinary matter and radiation very early on, gravity was much more easily able to retard and halt the expansion of localised regions where the density of cold dark matter was marginally greater than average. This caused the particles to fall together (see page 107), to form clumps, or 'haloes', which provided the gravitational wells within which ordinary baryons subsequently accumulated to create the first stars and galaxies. The smallest density fluctuations would have collapsed first, larger ones later. The eventual sizes of the dark matter haloes were determined by their masses (because CDM particles do not emit electromagnetic radiation, and scarcely ever collide and interact with other particles, they cannot get rid of the kinetic energy they acquire as they fall together; consequently, a CDM halo will cease to contract when its constituent particles are zooming around, like bees in a swarm, fast enough to resist the inward pull of gravity).

Recent calculations and simulations suggest that the first CDM haloes to form may have had masses of as little as one millionth of a solar mass (see Chapter 7). The precursors

of galaxies are likely to have been haloes with masses in the region of tens of thousands to tens of millions of solar masses (comparable to globular clusters and dwarf galaxies), and the growth of structure in the universe would proceed in hierarchical fashion through successive mergers of these haloes and the clumps of baryonic matter that were embedded within them. As galaxies and clusters formed and evolved, the relatively slow-moving cold dark matter particles would remain trapped within the merged haloes that surround the individual galaxies and clusters that we see in the universe today.

According to the hierarchical theory of galaxy formation small galaxies and their dark matter haloes merge successively to form larger galaxies. The reducing number of short arrows indicates the dwindling supply of gas that is falling into the haloes. Mergers between disc-shaped or spiral galaxies lead to the formation of massive ellipticals.

A universe filled with WIMPs

The fertile field of theoretical particle physics

Perhaps dark matter is not so cold

A recent study of a dozen of the dwarf spheroidal galaxies that surround the Milky Way, carried out by a team of astronomers from Cambridge, UK, together with collaborators from Hawaii and Switzerland, and published in 2006, has produced results that appear to throw the issue of 'warm' dark matter (particles that swarm around more slowly than hot dark matter particles such as lightweight primordial neutrinos, but faster than cold dark matter particles such as heavy WIMPs) into the frame.

With masses ranging upwards from just about ten times those of the most massive globular clusters, dwarf spheroidals are amongst the smallest and least massive galaxies of all. The Cambridge-led team used some of the world's largest telescopes to measure the spread of stellar velocities (the 'velocity dispersion') within each of the selected galaxies, at various distances between its centre and its outer fringes, and from these results were able to calculate the mass of its dark matter halo and the way in which dark matter is distributed within it. Remarkably, they found that the higher the luminosity (and hence the greater the amount of visible mass) of a galaxy, the lower the relative proportion of dark matter in its overall make-up. They concluded that each dwarf spheroidal, regardless of its overall size or luminosity, contains almost exactly the same amount of dark matter.

The observations also indicate that the smallest volume into which this amount of dark matter can be compressed has a diameter of about 1000 light-years. According to Gerry Gilmore, a member of the Cambridge team, the existence of dark matter haloes with these sizes and masses tells us a great deal about the nature of the dark matter particles. If their velocities were too high (i.e. if the dark matter were too 'hot'), gravity would be unable to hold these haloes together. On the other hand, if their velocities were too low (if the dark matter were too 'cold'), the haloes could be compressed to smaller sizes, and less massive haloes would be able to hold themselves together. The measured sizes and masses of the dwarf spheroidal haloes indicate that dark matter particles must be moving around substantially faster than standard cold dark matter models suggest. If this interpretation is correct, the dark matter which comprises the haloes of dwarf spheroidals appears to be 'warm', rather than cold[1].

If so, Gerry Gilmore argues that dark matter in those haloes is most likely to consist of low-mass WIMPs which, while hardly ever interacting with ordinary matter, do interact strongly with themselves. If the individual masses of the WIMPs were too high, there would be too few of them in each cubic centimetre for there to be enough interaction between them to sustain the relatively high spread of velocities (around 9 kilometres per second) that the observations imply. This suggests a need for much larger numbers of WIMPs with individual masses considerably less than the mass of a proton, and a lot less than the conventionally-assumed values of 1–1000 GeV/c^2. While it would be premature to draw firm conclusions about the nature of dark matter from one particular astronomical investigation, it is intriguing to see how observations of nearby 'rather boring' dwarf spheroidal galaxies may have the potential to put important constraints on the nature of the long-sought dark matter particles.

has thrown up a wide range of potential cold dark matter candidates, in particular various so-called WIMPs (Weakly-Interacting Massive Particles) that could have formed out of the extremely energetic radiation that pervaded the universe during its earliest instants. WIMPs are hypothesised particles, with masses ranging from a few to a few thousand GeV/c^2 (a few to a few thousand times the mass of the proton), which interact extremely weakly, and hence exceedingly rarely, with baryonic matter or electromagnetic radiation. The existence of particles of this kind is predicted by theories

that attempt to extend the Standard Model, and to unify the forces of nature. In particular, a hypothesis called Supersymmetry (SUSY) predicts that at very high energies – far in excess of what is found in the present-day universe, but which would have prevailed early in the history of the Big Bang – each fermion (matter particle) and boson (force-carrying particle) has an associated supersymmetric partner. For example, the SUSY partner of the photon (the force-carrying particle of the electromagnetic force) is the photino – a particle with a postulated mass in the region of 10–100 times that of the proton.

Matter fermions (quarks and leptons) have boson superpartners (squarks and sleptons), whereas force-carrying bosons (photons, W and Z bosons, gluons and Higgs) have fermion partners (with the suffix 'ino' – photino, W-ino, Z-ino, gluino, Higgsino). At one fell swoop Supersymmetry doubles the number of particles. In most Supersymmetry models, the lightest supersymmetric particle is stable and long-lived, which is what is needed if a particle of this kind is to be a candidate WIMP that could have been produced in the Big Bang, but still be around today. Within Supersymmetry theory, the lightest supersymmetric particle is the neutralino, a particle which is not itself an exact partner of any particular Standard Model particle, but which is formed when photinos, Higgsinos and Z-inos mix together to produce composite particles. The neutralino is currently the front-running WIMP candidate.

As with any other species of particle that existed in the hot dense fireball stage of the Big Bang (see Chapter 2), WIMPs would have been forming and annihilating (though WIMP-anti-WIMP collisions) with great rapidity, and their number density would at first have been comparable with the number density of photons. When the temperature dropped below the level at which WIMPs could be created, annihilation would have continued until eventually the chance of one WIMP finding another with which to annihilate became so small that their numbers stabilised. As the universe continued to expand, WIMPs became more dilutely spread through the expanding volume of space. Of the population of exotic particles that existed when the universe was very young, the more massive ones would long since have decayed, but less massive ones, such as the lightest neutralino would have been stable, and should still exist in large numbers. The number density of relic WIMPs in the universe today should be roughly inversely proportional to the strength of their interactions, and in order to provide the known density of dark matter in the present-day universe, it turns out that their interaction strength would have to be comparable to the rather feeble electroweak force (the unified weak and electromagnetic force) – hence the 'W' in WIMP. It is strikingly curious that SUSY theory, which was devised for completely unrelated reasons – in an attempt to solve problems with the Standard Model of particle physics – should have thrown up a range of particles with the right kinds of masses and interaction strengths to account rather neatly for the amount of cold dark matter that current cosmological measurements require.

The list of candidate particles emerging from theoretical attempts to extend the Standard Model, and develop an all-inclusive 'theory of everything' is long, diverse and intriguing; potential candidates include, in addition to the neutralino, the photino (superpartner of the photon) and the gravitino (superpartner of the graviton – the hypothesised force-carrying particle of gravity). More exotic possibilities include Kaluza-Klein (KK) particles, the existence of which is predicted if our four-dimensional space-time is embedded in a higher dimensional space (a concept which has its origins in ideas first suggested in the 1940s by Theodore Kaluza and Oscar Klein, who attempted to link gravity and electromagnetism by assuming the existence of a hidden extra dimension). If such theories are valid, the lightest Kaluza-Klein particle would be stable and would weigh in at around 200–1200 GeV/c^2 (about 200 to 1200 times the mass of the proton). Even more extreme is the Wimpzilla, a supermassive particle which, if it exists, would have a mass in the region of 10^{13} GeV/c^2 (ten trillion times the mass of the proton!).

Another candidate that has been canvassed is the axion, a type of particle that may have been formed during the transition from unified forces to separate electroweak and strong forces, a microscopic fraction of a second after the beginning of time. Having been produced in this way, these particles would have been 'cold' despite having very low masses. The axion would have zero charge, zero spin, and a mass which probably is considerably less than about 0.01 eV/c^2 (less than a hundred billionth of the mass of a proton). If indeed they do exist, axions – despite their exceedingly low

Shown here are some of the first magnets for the Large Hadron Collider installed, but unconnected, in the 27 km tunnel around which particles will be accelerated in order to collide with immense energies.

The path of the LHC tunnel is depicted below, superimposed on an aerial view of the environs of the CERN laboratory.

masses – would be so abundant that they could be viable cold dark matter candidates.

Just to show that neutrinos are not completely out of the frame, another outside bet is a hypothesised fourth type of neutrino, the 'sterile neutrino', or 'neuterino', which has no interactions with Standard Model particles apart from taking part in neutrino flavour mixing. If this particle exists (and there are only very marginal experimental hints that this might be the case), in order to avoid the problem of erasing small-scale structures in the early universe, its mass would have to be greater than 10 keV/c^2 (which is more than ten thousand times greater than the current

upper limits on the masses of the three known neutrino species).

Out of the veritable zoo of potential dark matter particles, the one which is currently most strongly favoured is the neutralino. So far, though, none of the proposed supersymmetric particles has been detected, but there are hopes that this might change within the next few years. In 2007, an immensely powerful particle accelerator, called the Large Hadron Collider, is scheduled to come on stream at CERN. Constructed in the same 27 kilometre circumference tunnel as an earlier accelerator (LEP), which was used to collide electrons and positrons, this machine will produce two beams of protons that will collide head-on with energies of around 14 TeV (14 million million electron volts), which is equivalent to about 15,000 times the rest-mass-energy of a proton. It will also be used to collide heavy ions such as lead with energies up to 1250 TeV. Collisions between particles at these huge energies will spray out all kinds of by-products, including, potentially, supersymmetric particles such as the neutralino, and the long-sought Higgs boson. Should particles of this kind be detected, it will greatly strengthen the case for SUSY particles, such as the neutralino, being a major dark matter constituent.

It seems, then, that most of the matter content of the universe is non-baryonic – matter, but not as we know it. MACHOs and neutrinos appear to be out (at most, they are minor players in the dark matter scene) and WIMPs are in, with the neutralino as the front-running candidate. However, before looking at the ongoing search for WIMPs (Chapter 7), it is worth taking a little time out to look at some niggling worries. While it is clear that cold dark matter models reproduce cosmic structures better than hot dark matter models, they are not without problems of their own – problems which have prompted some theoreticians to question whether dark matter really exists at all.

The Challenge of MOND - Does Dark Matter Exist at All?

According to the cold dark matter scenario, the formation of structure in the universe proceeds from the bottom up. Small spheroidal clumps, or 'haloes' of dark matter form first, within which baryons accumulate and fall together to create the first stars and protogalaxies (precursors of galaxies). These small dark matter haloes, and the baryonic matter that is trapped within them, grow together through a succession of mergers, to give rise to successively larger structures: small galaxies, then larger galaxies which, in turn, aggregate into the clusters, superclusters, sheets, filaments and voids that we see in the universe today. The ongoing growth of galaxies through collisions and mergers – with more massive galaxies cannibalising, disrupting and absorbing smaller ones – is readily apparent in the present-day universe.

Very large computer simulations are now able to study not only the general overall pattern of how structures grow in a universe where cold dark matter (CDM) is the dominant matter constituent, but can even study the formation and properties of individual galaxies. The results are fascinating and beautiful to behold. The simulations show vividly how both cold dark matter and baryonic matter accumulate to form a kind of cosmic web – rather like a spider's web, or a monstrous fishing net – within which matter flows along the filamentary strands of the net into concentrated blobs

This computer simulation shows the predicted distribution of dark matter in the universe. Galaxies form where denser concentrations occur within the complex network of filamentary dark matter structures. The region of space shown here is a billion light-years across.

NGC 1300 is a typical barred spiral galaxy. Spiral arms emerge from the ends of a straight 'bar', composed of stars, gas and dust, which straddles this galaxy's nucleus.

which become galaxies and clusters.

However, the cold dark matter scenario is not without its problems. First, models of CDM haloes predict that there should be a sharp rise in dark matter density – a so-called density 'cusp' – at their centres. The gravitational influence of so great a concentration of dark matter in its core would affect the orbital motions of stars and gas clouds in the central regions of galaxies and should, therefore, show up in their measured rotation curves. But as yet, except in a very few cases, astronomers have failed to find clear-cut evidence that the predicted density cusps are actually there. On the contrary, in many cases, there is evidence which appears to suggest that they are not. But the nuclei of bright spiral galaxies in many ways are not the best places in which to look for dark matter density cusps. Because so much of a spiral galaxy's luminous baryonic mass is concentrated in the central nucleus, the influence of any dark matter cusp on the inner part of its rotation curve is likely to be small. Furthermore, over aeons of time, gravitational

interactions between baryons and cold dark matter particles in a spiral nucleus may have redistributed the dark matter and flattened out any initial cusp in its density. This is likely to be especially so if, as in the case of the Milky Way Galaxy, the centre of the galaxy is straddled by a rapidly-rotating 'bar' of material which could catapult matter outwards from the centre.

As Michael Merrifield, of the University of Nottingham, UK, has pointed out, a better place in which to look for dark matter cusps is in 'low surface brightness galaxies', rather faint galaxies which have a very low density of visible matter in their central regions. Because of this, the gravitational influence of cold dark matter should determine the behaviour of their rotation curves close to the centre as well as further away, and the small amount of luminous (baryonic) matter should not have been able, over time, to redistribute the dominant dark matter. Therefore, the central cusp should still be present. But observa-tions of some low-surface brightness galaxies (notably NGC 6822, a nearby dwarf irregular

The low surface brightness galaxy, NGC 6822, is a member of the Local Group of galaxies and is classified as a dwarf irregular. In this three-colour image (taken in red, green and blue light), glowing gas clouds show up as green and red patches.

galaxy which lies at a distance of about 1.6 million light-years from the Earth) show that, although their rotation curves flatten out at large distances – consistent with the presence of a dark halo – they do not rise as steeply in their central parts as would be expected if dark matter cusps were there.

Secondly, computer simulations of the successive mergers of small-scale haloes to produce galaxies, predict the existence of large amounts of sub-structure in the form of numerous small satellite galaxies around each large galaxy – far more than has as yet been observed. Perhaps these large numbers of small satellite systems are actually there, but are simply so faint – or even completely dark – that they have not yet been detected. The recent discovery of galaxies such as VIRGOHI21 (see Chapter 3) which appear to consist almost entirely of dark matter, hints that we may be beginning to uncover elements of the predicted population of small-scale galaxy companions, but for the moment, there remains a clear mismatch between what hierarchical galaxy formation theory and simulations predict, and what we actually see.

Thirdly, the hierarchical galaxy formation scenario predicts that the smallest galaxies should have formed first and the largest, most massive, galaxies and clusters last. Yet recent observations at large redshifts – looking back to within a couple of billion years after the Big Bang – seem to indicate that massive galaxies and clusters were already in place at times considerably earlier than the favoured cold dark matter hierarchical clustering paradigm predicts.

A fourth point is that the random motions of stars in the outer parts of some (but by no means all) elliptical galaxies appear to decline with increasing distance from their centres in a way which suggests that they contain little, if any, dark matter at all. As we saw in Chapter 3, other explanations are possible (for

Numerous smaller dark matter haloes can be seen in this computer simulation, by a University of Zurich team, of the expected distribution of dark matter surrounding the Milky Way.

This false-colour optical image of a remote cluster of galaxies is overlaid with contours showing x-ray emission by hot gas. The cluster contains reddish elliptical galaxies which are full of old red stars. Located at a distance of some 9 billion light-years, this massive cluster must have formed when the universe was less than a third of its present age.

example, the decline in the measured velocities of these stars may be apparent rather than real – a consequence of stars moving in highly elongated orbits; or these particular galaxies may have been stripped of their dark matter haloes during close encounters with other galaxies). Nevertheless, these observations pose a significant challenge for the dark matter scenario.

So there are issues which the dark matter model has to address: an apparent lack, or shortfall, of dark matter in some galaxies, little sign of the expected dark matter cusp at the centres of low surface-brightness galaxies, apparently not nearly as many small satellite systems around massive galaxies as there ought to be, and the formation of high-mass galaxies and clusters earlier in the history of the universe than theory predicts. Whereas all of these issues may eventually be resolved satisfactorily within the dark matter framework – as a result of better observational data and further development of the theory – there are enough problems with the cold dark matter picture to cause at least some astronomers to look at possible alternatives to the whole dark matter concept.

Tinkering with gravity

There are two obvious explanations for the large discrepancy between the directly observable luminous masses of galaxies and clusters and the masses implied by their rotation and internal motions: either these systems contain very large amounts of unseen dark matter, or

the Newtonian law of gravity breaks down on the scales of these objects. Whereas the former idea is the widely favoured paradigm of our time, as early as the 1960s some theoreticians began to suggest that instead of invoking a load of dark matter to provide the required extra gravitational influence, perhaps the flat rotation curves of galaxies could be explained by changing the law of gravity. One of the first suggestions was that at large distances, or beyond a certain particular distance, the strength of gravity might decrease more slowly than the Newtonian inverse square law implies, in which case the orbital speeds of distant bodies (such as stars and gas clouds in the outer regions of galaxies) would no longer decline in the same sort of way as the orbital speeds of the planets decrease with increasing distance from the Sun.

One major problem about relating any change in the law of gravity purely to distance is that the bigger the galaxy, the more its rotational motion would deviate from Newtonian law, and the larger its apparent mass discrepancy would become. This does not match well with the observations – there are plenty of examples of small galaxies with large amounts of 'missing mass' and, on the other hand, examples of large galaxies, where the shortfall in mass is small.

In 1983, Mordehai Milgrom, of the Weizmann Institute, Israel, proposed that, rather than distance, it is the acceleration experienced by orbiting bodies and particles that is the critical factor. According to Newton's first law of motion, a body will continue to travel in a straight line at a constant speed unless acted on by a force. If a force is applied to that body, it will cause it to accelerate – that is, to change its state of motion. As a result, its speed may increase (or decrease), or its direction may change, or the outcome may be a combination of both of these things (its speed and direction may change). A body or particle that is moving at a constant speed along a circular orbit around a massive central body experiences a constant force of attraction, directed towards the centre, which causes its direction of motion to change continuously. It is subject to a constant acceleration, called centripetal acceleration, directed towards the central point. A useful analogy is to think of whirling a tennis ball around your body, attached to the end of a piece of string. The tennis ball is maintained in a circular orbit, and is therefore subject to a constant acceleration which is continually changing its direction of motion, by the pull you are exerting on it. If you let go of the string, the ball will carry on in a straight line in the direction in which it was heading at the instant you released it. Likewise, the centripetal acceleration experienced by any orbiting body – be it the Moon in its orbit around the Earth or a star in the outer fringes of a galaxy – is a consequence of the gravitational attraction exerted by the mass around which it is travelling (the Earth, in the case of the Moon; or the total amount of mass contained within its orbit, in the case of a star in the outer parts of a galaxy).

Milgrom devised a theory, called Modified Newtonian Dynamics, or MOND for short, in which the way in which gravity behaves depends on whether the accelerations experienced by bodies are greater or less than a certain particular transition value (denoted by a_0)[1]. In situations where accelerations are substantially greater than the transition value, the force of gravity behaves in the familiar Newtonian fashion (the force of gravity is inversely proportional to distance squared, so that if the distance is doubled, the force reduces to a quarter of its initial value). But where the accelerations experienced by bodies and particles are substantially less than the transition value, the force is inversely proportional to distance; in that regime, if the distance were doubled, the force of gravity would be reduced to a half (rather than a quarter) of its value. In the 'Newtonian regime', where gravity is an inverse square force, the speeds of orbiting bodies are inversely proportional to the square root of distance so that, if one body is four times further away than another, its orbital speed will be half that of the inner one; this is the familiar Keplerian pattern of rotation that we see in the Solar System. In the 'MOND regime', where the force of gravity is inversely

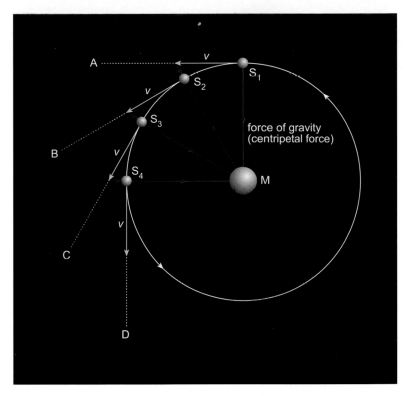

proportional to distance, orbital speed remains constant, regardless of distance, which is just what is seen in the outer parts of spiral galaxies where the observed rotation curves are flat (the speed remains the same regardless of distance).

Acceleration can be expressed as the amount by which a body's velocity (expressed in metres per second) changes in one second, and so is written as metres per second per second. Close to the surface of the Earth, the acceleration due to gravity (the rate at which the speed of a falling body increases) is about 9.8 metres per second per second. If you were to leap off the edge of a cliff then (ignoring the effects of air resistance) you would find yourself falling at a speed of 9.8 metres per second (35 kilometres per hour) after one second, 19.6 metres per second (70 kilometres per hour) after two seconds, and so on. According to MOND the transition acceleration has a value of around 10^{-10} metres per second per second, which is about one hundred billionth of the acceleration due to gravity at the surface of the Earth. Because the accelerations experienced by stars and gas clouds in the outer parts of galaxies are less than this value,

When a satellite that is moving round a massive body (M) at a constant speed (V) in a circular orbit is at position S_1, it is heading in direction A. In the absence of any forces, the satellite would continue in a straight line towards A, but the gravitational attraction of M deflects its path, causing it to accelerate, or 'fall,' in the direction towards M so that is successively reaches points S_2, S_3, S_4, and so on. It experiences a constant acceleration, which changes its direction but not its speed.

Milgrom's theory, if valid, places them firmly in the MOND regime and neatly accounts for why the rotation curves of spiral galaxies flatten out to constant orbital speeds at large distances from their centres.

Robert H Sanders, of the Kapteyn Astronomical Institute in the Netherlands, and Stacy S McGaugh, of the University of Maryland in the USA, and various other astronomers, have found that if they assume that the distribution of mass is the same as the observed distribution of luminous matter, then MOND gives remarkably good fits to the actual rotation curves of most of the spiral galaxies that they investigated. Moreover, these fits cannot be bettered by 'conventional' dark matter halo models. In addition to giving very good fits to the overall shapes of spiral galaxy rotation curves, MOND also reproduces individual detailed features (bumps and dips) in many of these curves when Milgrom's formula is applied to localised areas of enhanced brightness in these galaxies (areas where there is a greater concentration of luminous matter). The results are impressive. In light of these considerable successes, proponents of MOND contend that the dynamics of galaxies can be completely accounted for by luminous baryonic matter alone, without any need for dark matter.

MOND also neatly, and very precisely, accounts for a well-established observational result, known as the Tully-Fisher relationship, according to which the luminosities of spiral galaxies are proportional to the fourth power of the orbital velocities of gas clouds at their outer fringes. If all of a galaxy's mass were contained in its visible stars and gas clouds (i.e., if it did not have a dark matter halo) then its luminosity would be directly proportional to its mass – the greater its mass, the more luminous it would be. In those circumstances, according to the conventional Newtonian theory of gravitation, the square of the velocity at the outer perimeter of a spiral galaxy would be proportional to its mass and hence to its luminosity. But according to MOND, the fourth power of the orbital velocity is directly proportional to mass. Consequently, in the absence of dark matter, the theory naturally predicts that the luminosity of a spiral galaxy likewise should be directly proportional to the fourth power of its rotational velocity – exactly as the Tully-Fisher relationship requires. It is much more difficult, and more 'messy' to account for the Tully-Fisher relationship with dark matter haloes.

Without any need to invoke dark matter, MOND provides good fits to the rotation curves of at least some of the low surface brightness galaxies. Furthermore, according to proponents of MOND, it comes as no surprise that stellar velocities (as measured from planetary nebulae) in at least some small elliptical galaxies appear to decrease with increasing distance from the centre. They argue that because elliptical galaxies are much more concentrated than spiral ones, the gravitational accelerations at their visible boundaries are higher than the MOND transition value and so ordinary Newtonian dynamics should apply – leading naturally, in the absence of dark matter, to orbital velocities that decrease with distance.

On the face of it, MOND seems to have quite a lot going for it – it explains away the flat rotation curves of spiral galaxies, is consistent with the declining rotation curves of some elliptical galaxies, and rather neatly accounts for the Tully-Fisher relationship. In a number of respects it seems to fit the observational data on individual galaxies just as well – perhaps even better – than the dark matter alternative.

It appears to be less successful, though, in dealing with clusters of galaxies. When the overall masses of galaxy clusters are computed using MOND, they turn out to be lower than the masses calculated using Newtonian dynamics but still nearly twice as high as the total amount of mass that has actually been detected in the form of x-ray emitting gas and the luminous stars that comprise their constituent galaxies. Indeed, a recent investigation using high quality x-ray data from the XMM-Newton satellite, carried out by Etienne Pointecouteau and Joseph Silk of the University of Oxford, UK, suggests that in the outer regions of clusters the discrepancy between the MOND mass and the baryonic mass may be a great as a factor of four or five, which would imply that, even under MOND, about 80 percent of the mass in these clusters is 'missing'. Even MOND still seems to require dark matter of some kind (dark baryons, neutrinos or something else) to account for a substantial proportion of the overall masses of galaxy clusters.

Riccardo Scarpa, of the European Southern Observatory, has suggested that so-called ultra-compact galaxies (UCDs) may provide a good test bed for the predictions of MOND. These objects, which have properties intermediate between those of globular clusters and dwarf galaxies, were discovered in 1999 in a survey of the Fornax galaxy cluster (which lies at a distance of some 60 million light-years), and more have recently been found in the similarly-distant Virgo cluster. A distinctive feature of these tiny galaxies, which have masses in the region of 10–100 million solar masses and radii in the region of 30–100 light-years, is that stars are strongly concentrated towards their centres. In the UCDs that have been investigated, the mass-to-light ratios are only in the range 2–4, which closely matches what would be expected for a population of old stars with little or no dark matter. The apparent absence of dark matter could simply be due to these objects being the end products of the merging of giant star clusters formed during periods of strong galaxy interactions. Another possibility is that they represent the nuclei of normal dwarf galaxies that have lost, or been stripped of, their external halo and dark matter. But a third possibility is that the apparent absence of any 'missing mass' ties in with the fact that in all of these objects the outer edge acceleration is still well above the level at which the MOND regime would kick in. In the absence of dark matter, velocities should decline with increasing distance. As Scarpa has pointed out, if even one of these objects were to be shown to have an incontrovertible large mass discrepancy, it would falsify the MOND theory.

As it is, there are objects such as VIRGOHI21 that seem to pose problems for MOND in any case. VIRGOHI21, which contains no stars but does contain a hundred million Suns' worth of hydrogen gas, appears from its overall dynam-

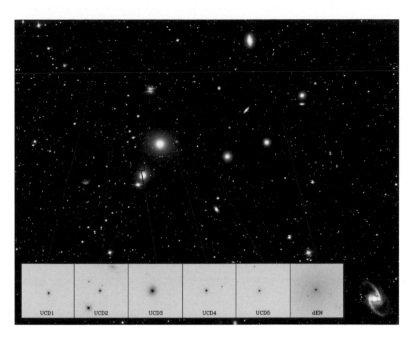

ics to have a total mass of about 10 billion Suns – in other words, to be composed almost entirely of dark matter. However, the accelerations experienced by gas clouds at the outer detectable limits of this object are far in excess (by a factor of a hundred) of what MOND would predict if the detectable matter (the hydrogen gas) were the sole constituent of this object. In other words, if the observed properties of this galaxy are correct, dark matter appears to be essential to explain its dynamics – the MOND hypothesis fails to do so. A similar problem faces MOND on the much smaller scale of open star clusters, such as the Pleiades. The accelerations experienced by stars in the outer parts of these clusters are well below the MOND transition, but there is no evidence at all of apparent missing mass in clusters of this kind. On the other hand, Milgrom has argued that the behaviour of modest open clusters such as the Pleiades is dominated by the influence of the gravity of the Galaxy as a whole.

Critics argue that MOND is a wholly ad hoc theory – devised specifically to explain away the observed flat rotation curves of galaxies. Rather like 'single issue politics', there was no other compelling physical reason for inventing it. Furthermore, if MOND is a valid concept, then it involves a violation of Newton's law of universal gravitation, which implies that

The lines in the main image of the Fornax cluster show the locations of the ultra-compact dwarf galaxies that are featured in the negative-print boxes at the bottom. Located at a distance of some 60 million light-years, these tiny galaxies are only about 120 light-years across.

Einstein's general relativity would also be violated. Consequently, until recently, a further criticism of MOND was that no one had developed a formulation of MOND based on relativistic principles. That situation changed in 2004 when Jacob D Bekenstein of the Hebrew University of Jerusalem constructed a theory called TeVeS (for 'tensor-vector-scalar field theory') that seems to fit the bill rather well: it behaves like Newtonian dynamics at low velocities (just as relativity does) and behaves like MOND at low accelerations. Another criticism of MOND was that no-one had shown whether, or how, MOND could explain the strong gravitational lensing effects produced by galaxy clusters which appear to be far in excess of what luminous matter alone could produce within conventional gravitation theory. Bekenstein has demonstrated that TeVeS works for photons as well as particles of matter, and has shown that it produces gravitational lensing effects very similar to those that have been predicted by dark matter models.

An immense amount of work will need to be put in to find out whether or not MOND can reproduce the full panoply of cosmological and astrophysical phenomena which conventional gravitational theory, and the cold dark matter hypothesis, appear to deal with rather well on the whole. Unless and until it can get to that stage, MOND will always remain an outside bet compared to the widely favoured dark matter alternative. Some theoreticians have already begun to explore these possibilities. For example, in the summer of 2005, a team headed by Constantinos Skordis of the University of Oxford showed that it may be possible for relativistic MOND to reproduce the observed features of the cosmic microwave background and galaxy distributions, but only if neutrinos make a substantially greater contribution (equivalent to about 15 percent of the critical density) to the overall density of the universe than current estimates suggest.

MOND's genesis may lie in an ad hoc attempt to explain away the flat rotation curves of spiral galaxies but, as its supporters would argue, is it any more arbitrary to try to resolve the problem by proposing a modification to the law of gravity than by invoking the existence of invisible mass of an unknown nature, which has not been detected in tangible form after more than 30 years of trying? As we have seen, MOND has problems and challenges to contend with – not least its inability to account fully for the 'missing mass' in galaxy clusters. But with its considerable apparent success in describing the dynamics of rotating galaxies, coupled with the added depth of Bekenstein's relativistic formulation, MOND deserves to be taken seriously, and cannot be dismissed out of hand

Nevertheless, while MOND cautions us to keep our minds open to the possibility that dark matter may not exist at all, in the face of such a large and varied body of evidence pointing in the same direction – rotation curves, dynamics of clusters, gravitational lensing by clusters, weak lensing, detailed analyses of the cosmic microwave background, and galaxy distribution surveys, together with new evidence of what appear to be dark galaxies, and the successes (despite still unresolved difficulties) of CDM simulations and the hierarchical galaxy formation model – most cosmologists remain convinced that dark matter really does exist, and outweighs ordinary matter many times over.

But as Michael Merrifield, of the University of Nottingham, has remarked, 'Direct laboratory detection of massive particles from the halo of the Milky Way would provide the most convincing confirmation of the whole dark matter paradigm, and would lay the issue to rest once and for all'.[2] This challenge has been taken up by about twenty different groups of researchers across the globe – the WIMP Hunters.

The WIMP Hunters

WIMPs are the front-running candidates for being the cold dark matter particles that are widely believed to comprise about 22 percent of the total mass-energy content of the universe. Should new generations of high-energy particle accelerators, such as the Large Hadron Collider (LHC), succeed in creating and detecting WIMPs, that would be a spectacular leap forward for particle physics. Such a discovery, of itself, would not prove that WIMPs are the major dark matter constituent of the cosmos, though it would render the hypothesis much more plausible. But if experimenters were unequivocally to detect WIMPs originating from the halo of our Galaxy, that would be a profoundly significant discovery not only for cosmology but also for particle physics. For the cosmologist, the solution to the cosmic dark matter problem would be at hand. For the particle physicist, the detection of WIMPs from the galactic halo would provide clear evidence that some of the particles predicted by extensions of the Standard Model of particle physics, such as Supersymmetry (SUSY), do indeed exist in nature. The importance of this quest is epitomised by the fact that more than twenty different research teams around the globe have experiments up and running, or currently being developed, to track down and identify WIMPs of cosmic origin.

The WIMP hunters have two principal lines of attack – direct detection experiments, and indirect detection experiments. Direct detection experiments attempt to detect the effects produced on those rare occasions when a WIMP collides with an atomic nucleus inside a detector, whereas indirect detection involves scanning the sky for the decay products – such as bursts of gamma-rays – which are produced

when large numbers of WIMPs collide and annihilate each other.

Direct detection presents an enormous challenge to experimental physics. Because they interact so exceedingly weakly with matter, the chance of an incoming WIMP colliding with an atomic nucleus within a detector is extremely low – current estimates suggest there would probably be less than one collision per day per 10 kilograms of detector material. To put this another way, of the tens of trillions of galactic halo WIMPs that theory suggests must be passing through your body every day, only the odd one or two are likely to collide with any of the more than 10^{28} baryons of which you are composed. Nevertheless, the basic idea for WIMP detection is very simple in principle. If a massive WIMP, probably weighing between about ten and a thousand times the mass of a proton (i.e. 10–1000 GeV/c^2), strikes an atomic nucleus, the nucleus will recoil under the blow. The kinetic energy (energy of motion) of the recoiling nucleus is expected to be very small by particle physics' standards (in the region of a few thousand to a few tens of thousands of electron volts), but should, nevertheless, be detectable.

The probability of a WIMP interacting with a nucleon (a proton or neutron in an atomic nucleus) is usually expressed in terms of a quantity called the WIMP-nucleon cross-section, which, in effect, is the target area that a nucleon presents to an incoming WIMP. Think of the WIMP as a bullet and the nucleon as the bull's eye on a shooting gallery. The likelihood of a hit depends on the cross-sectional area of the bull's eye – the bigger the bull's eye, the greater the chance of a hit. Nuclear physicists have devised a unit of their own to describe the tiny cross-sections of particle interactions. Called a barn, it is equivalent to 10^{-24} centimetres squared – one million, million, million millionth of a square centimetre – and is roughly equal to the cross-sectional area of an atomic nucleus (particles have to hit a very tiny 'barn door' indeed!). Yet even a barn is cumbersomely large for WIMP-nucleon cross-sections. Instead, particle physicists talk in terms of picobarns, where 1 picobarn (symbol pb) is a trillionth (i.e. a million millionth) of a barn, or 10^{-36} centimetres squared. For WIMP-nucleon interactions, the cross-sections are believed to be very tiny indeed, almost certainly considerably smaller than 10^{-42} centimetres squared, or 10^{-6} picobarns (one millionth of a picobarn). This corresponds to an effective target area at least a million million million (10^{18}) times smaller than the cross-sectional area of an atomic nucleus. Since predicted cross-sections for neutralinos – the most popular WIMP candidates – are in the region of 10^{-6} to 10^{-12} picobarns, detecting them presents a monumental challenge.

Current generations of experiments attempt to do so in one or more of the following ways: by detecting light that is emitted when a nucleus recoils (scintillation), by measuring the ionisation produced when a nucleus recoils (the recoil knocks electrons off atoms in the detector, which can then be detected and measured) or by measuring the microscopic quantity of heat dumped into the detector by the collision. This heat manifests itself in what are known as phonons, tiny bundles, or quanta, of vibrational energy that are deposited into a detector crystal (just as a photon is a quantum of electromagnetic energy, so a phonon is a quantum of vibrational, or sound, energy).

In addition to the technological challenge of building detectors that are sufficiently sensitive to detect these elusive particles, the WIMP hunters have another major problem to contend with: for every (possible) WIMP interaction there are millions of unwanted background events. Detectors have to be exceedingly efficiently shielded in order to screen out, or at least minimise, extraneous events which could mimic the effects of a WIMP collision. Major culprits include cosmic rays (highly-energetic particles, such as electrons, protons, and atomic nuclei, which travel through space at very large fractions of the speed of light); natural radioactivity in terrestrial rocks (which produce electrons, neutrons, alpha particles, and gamma-rays in far greater abundance than

This view shows the external laboratory buildings of the Gran Sasso Underground Laboratory, Italy.

expected numbers of WIMP events); and particles and radiation originating in the electronics of the detectors. All materials in everyday use, and ourselves for that matter, contain minute quantities of naturally radioactive substances; the materials used to construct WIMP detectors and their associated electronics have to be very carefully selected and purified to reduce radioactive impurities to an absolute minimum.

The first crucial step is to locate detectors deep underground, so that the overlying rock absorbs and drastically reduces the effect of cosmic rays and their by-products. A kilometre or more of overlying rock can diminish the cosmic ray problem by a factor of about a million. The detector must then be surrounded by a shield, or concentric shields, of material such as water, wax, copper, lead or polyethylene to absorb neutrons, electrons and gamma-rays from the surrounding rocks and, indeed, from the instrument's own electronics. Neutrons – which are produced by natural radioactivity or from collisions between cosmic-ray muons and atomic nuclei – are particularly troublesome because they cause nuclear recoils similar to those expected from WIMP interactions. Because the shields cannot stop everything (WIMPs, of course, will pass through them as

if they were not there), experimenters usually surround WIMP detectors with devices that record the influx of the remaining background so that the effects they produce inside the detector can better be identified and weeded out. Crucially, detection systems must be able to distinguish between nuclear recoils caused by WIMPs (and, unfortunately, by neutrons too) and electron recoils (caused by remaining background electrons, x-rays and gamma-rays).

One of the longest-running subterranean research programmes, which began in 1990, is DAMA (DArk MAtter) – a joint Italian-Chinese project housed at a depth of 1400 metres in the Gran Sasso Underground Laboratory near Rome. One of DAMA's key instruments was a 100-kilogram sodium-iodide (NaI) scintillator (replaced in 2003 by a 250-kilogram version), which recorded flashes of light emitted by events taking place within the detector. The basic idea is that when a nucleus recoils, it excites electrons in neighbouring atoms to higher energy levels; as they then drop back down to lower levels, they emit light (see Chapter 1). The likelihood of a WIMP interaction depends on the mass and speed of the incoming WIMP and the mass of the target nucleus, and is highest when the mass of the

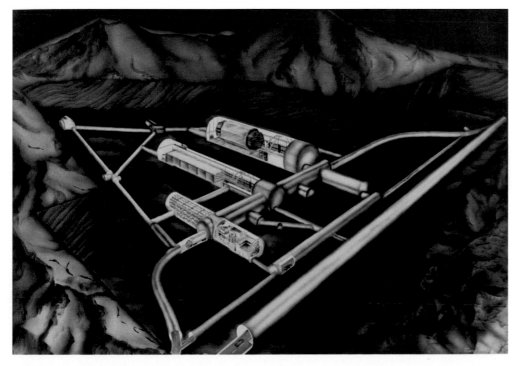

This artist's impression shows the Gran Sasso Underground Laboratories, which are located at a depth of 1400 metres below ground and which house various experiments, including DAMA.

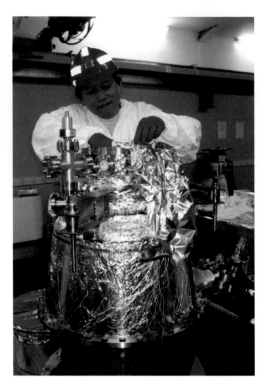

recoil events. During its period of operation, from 2001 to 2004, ZEPLIN I did not record any WIMP-induced nuclear recoils, but from their data, the experimenters were able to show that the WIMP-nucleon cross-section must be smaller than 1.1×10^{-6} pb.

One of the advantages of xenon detectors is the relative ease with which they can be scaled up to larger and more massive versions. ZEPLIN II, which was installed underground in 2005 contains 10 times as much liquid xenon as its predecessor, is better able to discriminate between nuclear and electron recoils, and has a higher sensitivity, which hopefully will push its detection limit down to less than 10^{-7} pb. Further improvement is expected with the next phase, ZEPLIN III. The ultimate goal is to build a 1-tonne detector, ZEPLIN-MAX, which ought to be able to achieve a sensitivity of around 10^{-9} pb. Liquid xenon is also the chosen medium for a series of detectors being developed by the XENON collaboration, headed by the University of Columbia, New York, which aims eventually to achieve a sensitivity in the region of 10^{-10} pb with a 1-tonne detector that it plans to construct in the Gran Sasso Underground Laboratory.

An alternative approach is to go to low-temperature (cryogenic) detectors. Devices of this kind use crystals of materials such as germanium or silicon, cooled to temperatures as low as 0.01 K (10 mK) – 0.01 degrees above Absolute Zero (Absolute Zero is the lowest possible temperature – the temperature at which all molecular motion ceases). At such extreme low temperatures, the small amount of energy deposited in the crystal by a WIMP–nucleon collision could change its temperature by a tiny, but measurable, amount (the anticipated temperature changes are around a hundred thousandth of a degree, which, at a detector temperature of around 0.01 K, corresponds to a change of about one part in a thousand). At these temperatures, the background of tiny vibrations in the crystal lattice (phonons) is exceedingly low, making it much easier to distinguish and count the phonons generated by nuclear recoils and hence to measure the (heat) energy deposited by these events. But

target nucleus is comparable to the energy of the WIMP. Because sodium iodide contains relatively light sodium atoms (atomic mass 23) and heavier iodine atoms (atomic mass 127) it should be sensitive, in principle, to a wide range of WIMP masses.

Boulby Mine, a working potash and salt mine on the edge of the North Yorkshire moors in the north-east of England, plays host to a range of experiments operated by the United Kingdom Dark Matter Consortium (UKDMC). At a depth of 1100 metres these experiments are located in caverns excavated out of the surrounding salt rock, a mineral that happily has a low level of natural radioactivity. One of the experiments – NAIAD – employs an array of sodium iodide detectors which operate on a similar principle to the DAMA NaI detectors. Another utilises liquid xenon (Xe) as a detecting material. The first of its developing series of xenon-based detectors was ZEPLIN I (the acronym derives from ZoneEd Proportional scintillation in LIquid Noble gases), the heart of which was a copper vessel containing 3.2 kilograms of liquid xenon that was viewed by three photomultipliers orientated so as to pick up the flashes of light that were produced by temperature measurements alone are not sufficient to separate nuclear recoils (due to WIMPs and neutrons) from electron recoils (due to the radioactive background). However, particle recoils also ionise neighbouring atoms, and in germanium or silicon the ionisation yield (the amount of electrical charge released in relation to recoil energy) differs significantly between electron recoil and nuclear recoil events. By simultaneously measuring ionisation and phonons, experimenters can establish whether an event is due to a genuine WIMP interaction, or to the unwanted background. This approach has been adopted by the American CDMS (Cryogenic Dark Matter Search) and the French EDELWEISS projects. A related project, CRESST, relies on simultaneously measuring ionisation and scintillation (nuclear and electron recoils generate different amounts of light) to identify nuclear recoil events.

The CDMS detectors consist of germanium or silicon discs, called ZIPs (Z-dependent Ionization Phonon detectors), which are each about the size of a thick chocolate biscuit, arranged in stacks, or 'towers'. Because the original experiment (CDMS I) was housed in a shallow underground tunnel underneath the Stanford University campus, California, background radiation and particle levels were high. The current set-up (CDMS II) is located in the defunct Soudan mine in Minnesota, at a depth of 780 metres. Early in 2004, the CDMS II team published the first results from 'Tower 1', a stack consisting of four germanium and two silicon ZIPs. Because they found no WIMP-induced recoils in 52.6 kilogram-days of data (the mass of the detector multiplied by the duration of the exposure), the experimenters were able to set an upper limit of 4×10^{-7} pb on the WIMP-nucleon cross-section for WIMP masses in the region of 60 GeV/c^2 – a figure which was four times lower than the most sensitive previous result (obtained by EDELWEISS in 2002) and eight times better

Clockwise from top left: A view of the inner layers of the CDMS II cryostat (cooled vessel) with two towers, each consisting of a stack of six ZIP detectors, installed. The towers are located towards the upper edge of the circular base.

One of the ZIP detectors, which is used in the Cryogenic Dark Matter Search (CDMS II) WIMP detector in Soudan Mine. The thin film on its surface is the phonon sensor.

A scientist examines the shielding around the CDMS II cryostat, the cooled container within which the experiment is housed and cooled to temperatures in the region of 0.02 degrees above Absolute Zero.

Located some 780 metres below the surface workings of this former iron mine in Soudan, Minnesota, is the underground laboratory that houses the CDMS II WIMP detector.

As the Sun travels round the centre of the Galaxy (a) it ploughs through the halo of WIMPs at a speed of about 230 kilometres per second. The Earth's motion around the Sun (b) gives rise to a seasonal variation (c) in the speed at which WIMPs appear to be 'blowing' past the Earth. (d) The speed of the apparent WIMP wind reaches a maximum in June and a minimum in December.

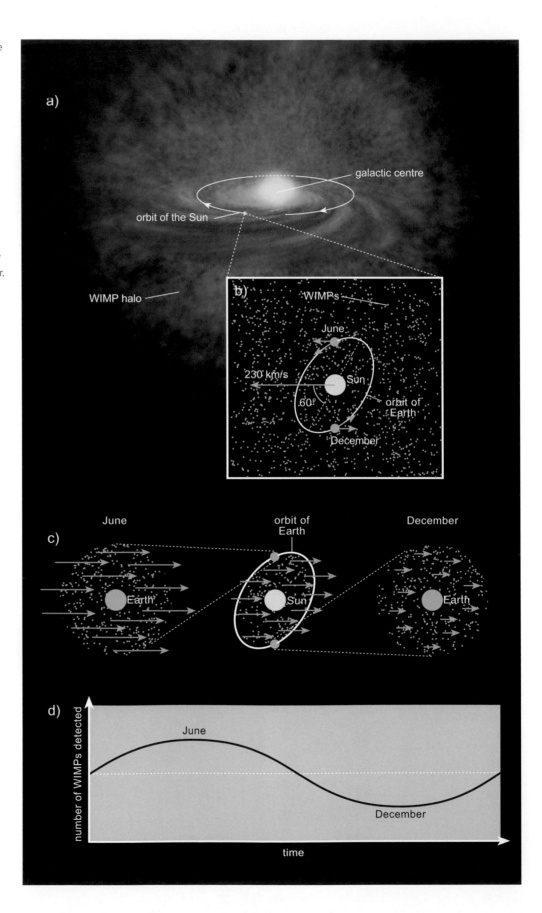

than CDMS I had achieved. By mid-2005, with two towers up and running, the CDMS II team were able to set the most stringent limits yet on the WIMP-nucleon cross-section – a figure of 1.6×10^{-7} pb. Put another way, these results imply that the interaction rate of WIMPs must be less than one event per 60 days per kilogram of germanium.

The EDELWEISS experiment is located at a depth of 1660 metres (more than twice the depth of Soudan) in the Laboratoire Souterrain de Modane (Modane Underground Laboratory) beneath the French-Italian Alps. By 2002, the EDELWEISS team had found an upper limit for WIMP-nucleon interactions of about 10^{-6} pb, a figure comparable to that obtained by ZEPLIN I, but which has since been trumped by the CDMS II results. Currently under construction, EDELWEISS II, which will have a cryostat (cooled container) capable of holding up to 120 detectors, is expected to become a hundred times more sensitive than the original EDELWEISS I experiment. Within the WIMP-search 'industry', detection limits of 10^{-7}–10^{-8} pb are likely to be achieved soon. The development of 1-tonne-class detectors, which may become operational and begin to produce useful data between 2008 and 2011 should push sensitivities down to 10^{-8}–10^{-10} pb. But to go as far as 10^{-12} pb would require 100-tonne detectors. Leading contenders to develop 1-tonne detectors include ZEPLIN-MAX, at Boulby, XENON1T at Gran Sasso, XMASS, at Kamioka, Japan, and Super CDMS, an ambitious proposal to develop a cooled assembly of germanium ZIP detectors with a total mass of 1 tonne.

If, or when, direct detectors finally succeed in detecting WIMPs, the next important step will be to show that they are actually coming from the galactic halo. The key to determining this will be to look for effects caused by the motion of the Earth through the galactic halo, either by measuring seasonal variations in the number of WIMPs detected or by detecting the predominant direction from which WIMPs are arriving and measuring changes in that direction caused by the rotation of the Earth around its axis (diurnal variations).

The WIMP wind

Imagine, for a moment, that you are a dedicated athlete running doggedly round a circular track despite the incessant onslaught of a hailstorm driven by a gale-force wind. While pounding around the circuit you will become acutely aware that the stinging impact of the hailstones is greatest when you are running directly into the wind, whereas when you get round to the far side of the track and are running downwind, the hits will be slower, fewer, and less painful. The frequency and energy of the impacts will rise and fall in a regular cycle, lap after lap, for as long as the hailstorm persists.

If our Galaxy's dark matter halo is indeed made up of WIMPs, the combined motion of the Sun and the Earth will cause WIMP detectors to 'suffer' in a similar way. The Sun travels along a near circular orbit around the centre of the Galaxy at a speed of about 230 kilometres per second (about 800,000 kilometres per hour), while the Earth moves around the Sun at about 30 kilometres per second (108,000 kilometres per hour). Consequently, from the Sun's perspective, WIMPs will appear to be streaming steadily past at an average speed of 230 kilometres per second in the opposite direction to the Sun's motion, like hailstones carried on a wind. The Earth's orbital motion causes the apparent speed and strength of the WIMP wind to vary throughout the year, reaching a maximum around the beginning of June (2 June) when the Earth is heading in the same direction as the Sun (upwind) and dropping to a minimum, six months later, when the Earth is heading in the opposite direction (downwind).

If the plane of the Earth's orbit coincided exactly with the plane of the Sun's orbit, the observed speed of the WIMP wind would increase and decrease by 30 kilometres per second – a variation of 13 percent. However, because the Earth's orbit is tilted at an angle of 60 degrees to the plane of the Sun's motion, only about half of our planet's orbital speed is added to, or subtracted from, the speed of the Sun, and the resulting annual variation in the WIMP wind is expected to be less than 7 percent. Taking into account the orientation of the

Earth's orbit, the number of events recorded by a WIMP detector should rise to a maximum in early June and drop to a minimum in December.

DAMA provokes a controversy

The DAMA programme, headed by R Bernabei of the Università di Roma, has been searching for the seasonal WIMP signature since the mid-1990s. Early in the year 2000, the DAMA team announced that, for three consecutive years, the event rate in their 100-kilogram sodium iodide scintillator had shown a small (~2 percent) seasonal variation which peaked in early June and dropped to a minimum in December. They contended that the varying event rate was caused by halo WIMPs weighing in at about sixty times the mass of a proton. Others were highly sceptical of this startling result, suggesting that other factors, such as seasonal variations in the detector, were responsible.

By July 2002, the DAMA researchers had accumulated seven years of data (adding up to 107,731 kilogram-days of detector exposure) which clearly showed an ongoing seasonal variation, and by the time the full analysis of the data was published in the summer of 2003, were more than ever convinced that the effect they had been measuring was real and that it was caused by galactic halo WIMPs with masses of 52 GeV/c^2 and cross-sections of 7.2×10^{-6} pb. Statistically, the results were highly significant, and the dates of the peaks and troughs matched well with the times in the year when the combined motion of Earth and Sun would maximise and minimise the WIMP wind. The results seem compelling, yet appear to be contradicted by several competing experiments – notably EDELWEISS, ZEPLIN I and CDMS II. None of these detectors has as yet recorded a single event which displays the clear-cut hallmark of a WIMP-induced recoil, but each experiment has set strict limits on the detectability of WIMPs that, on the face of it, appear to rule out the detections claimed by DAMA. In particular, the upper limits set by CDMS II in 2005 seem clearly to imply that the cross-section for WIMPs with a mass equivalent to about sixty proton masses must be at least

forty times smaller than the cross-section implied by the DAMA results, unless some other factor, such as particle spin (elementary particles can be visualised as tiny spinning balls) needs to be taken into account. But even then, the limits set by various different groups appear to exclude the DAMA results except for a very limited range of WIMP masses towards the lower end of the expected range.

The DAMA experimenters remain resolutely unfazed. They contend that the seasonal variation is independent of any particular theoretical model of the nature of WIMPs and how they interact with nucleons, and that other teams – using different kinds of detectors and different sets of assumptions – are not comparing like with like. One of the points which they make is that their observed modulation relates to single-hit events, in which just one detector out of many records evidence of a hit. Because of the extremely low probability of a WIMP–nucleon interaction occurring at all, the probability that a particular WIMP would scatter off more than one detector is utterly negligible. Their results for multiple-hit events (events caused by background) do not show the seasonal modulation. This, they assert, argues in favour of the measured modulation being due to galactic WIMPs, not a seasonal change in the background.

Has the DAMA team, whose sodium iodide detector has racked up more observing time than any other experiment, found genuine evidence of WIMPs in the galactic halo, or is the observed variation a spurious phenomenon induced by as yet unidentified factors that mimic the effects of a WIMP wind? The controversy is likely to rumble on unabated unless and until other research groups find incontrovertible evidence for the existence of WIMPs, or theoreticians find a way of reconciling the apparent contradictions between the DAMA results and the rest.

Checking out directions

Measuring recoil directions should provide a more powerful and decisive test for galactic WIMPs. Currently, the Sun's motion around the galactic centre is taking it towards a point in the constellation Cygnus. If, as the simplest halo models assume, the dark matter halo is a non-rotating sphere (the WIMPs of which it is composed zoom around in random directions rather like the constituent stars in globular clusters or elliptical galaxies, but the halo as a whole does not rotate in any particular direction), the WIMP wind should be blowing from the direction of Cygnus. The direction from which any particular WIMP arrives will depend on the combined effect of its individual motion within the halo plus the Sun's orbital motion, but the difference between the numbers arriving from the Cygnus direction and the numbers arriving from the opposite direction could be as much as 100 to 1. If the directions of a large number of WIMP-induced recoils are plotted on a graph they will show a strong peak in the direction directly opposite to Cygnus. Furthermore, as the Earth rotates on its axis, the mean recoil direction in the detector will change in a regular, periodic fashion as the daily rotation causes Cygnus to track across the sky.

At the latitude of Boulby, where the UKDMC team, together with international collaborators, are developing a detector to look for precisely these effects, the mean recoil direction is expected to rotate from downwards (into the ground) to southwards and back over a sidereal day. Because the sidereal day (the true rotation period of the Earth relative to the background stars) is about four minutes shorter than the mean solar day (the basis of everyday clock time), the timings of these variations will gradually and progressively get out of step with civil clock time at a rate of four minutes per day. Both of these effects – the periodic change in recoil direction and gradual change in the times at which the variations occur – are linked to the true rotation period of the Earth and cannot be mimicked by ordinary background events in the detector and its surroundings, or by other terrestrial sources. If detected, diurnal variations in mean recoil directions would provide a very strong 'signature' of the detection of galactic WIMPs, particularly if the mean recoil direction turned out to be 180 degrees away from the constel-

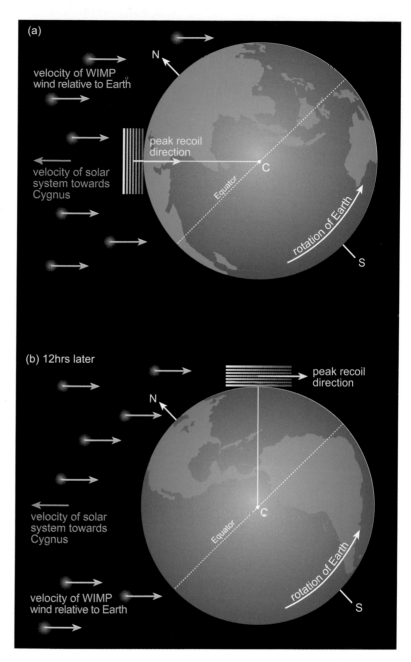

lation of Cygnus. Current estimates suggest that fewer than 100 directional WIMP events, perhaps even as few as 10, would be enough for experimenters to be at least 90 percent confident that they had detected a genuine galactic WIMP signal. This compares with the thousands of events that would be required to be similarly confident of having detected the annual modulation in the number of WIMPs caused by the Earth's orbital motion around the Sun.

The Boulby experiment is called DRIFT

The rotation of the Earth causes a cyclic daily change in the peak recoil direction of target nuclei in a WIMP detector (shown symbolically by the grid of parallel lines). At the particular latitude shown, the peak direction should vary from almost vertically downwards (a) to near-horizontal (b) and back in a period of 24 hours.

(Directional Recoil Identification From Tracks). Developed and built by Sheffield University on behalf of the UKDMC and collaborators from Occidental College, Los Angeles and the University of New Mexico, DRIFT is the first detector in the world with the capability of measuring the directions of WIMP-induced recoils. Following a WIMP impact, a recoiling nucleus will produce a trail of electrons and ions as it ploughs through the detector material. The basic idea behind DRIFT is to determine the track of the nucleus by measuring the trail of ionisation that it produces. An added benefit of a device of this kind is that it can distinguish very efficiently between genuine recoils and events produced by electrons and alpha particles by looking at their track lengths – electrons and alpha particles have much longer ranges than recoil nuclei. However, because WIMP-generated nuclear recoils are expected to have very low energies, they will travel only very short distances in solids, liquids or gases at everyday temperatures and pressures. To enable nuclear recoils to produce measurable trails (a few millimetres long), DRIFT uses low-pressure gas; the downside to this is that the detector mass is low (the cubic-metre volume of the DRIFT I prototype, which ran between 2001 and 2004, contained less than 0.2 kilograms of the selected target gas – carbon disulphide) and the likelihood

of WIMP-induced events is correspondingly reduced.

DRIFT II, which consists of an array of DRIFT modules, has been operational since March 2005. Because the instrument's target mass increases each time an extra module is added to the array, the experimenters expect DRIFT II eventually to achieve a 30- to 50-fold increase in sensitivity over the original prototype. The next generation instrument (DRIFT III) will consist of 100 modules and should have a potential sensitivity of around 10^{-8} pb. The ultimate goal would be a one-tonne directional detector with a sensitivity in the region of 10^{-10} pb.

Japanese experimenters are also developing a gaseous directional detector called NEWAGE. Currently at an early stage of development, this instrument, which will be based at the Kamioka laboratory, should eventually be able to measure nuclear recoils' tracks directly in three dimensions rather than having to reconstruct 3-dimensional tracks from two-dimensional readouts (as on the current version of DRIFT).

The standard, and simplest, assumption is that the galactic WIMP halo is a non-rotating sphere through which the Solar System ploughs at a speed of about 230 kilometres per second. In reality, the halo may have a different form, perhaps somewhat flattened (a spheroid) or even triaxial (having different diameters along three mutually perpendicular axes). To add to the possible complications, dark matter may be distributed in a clumpy fashion, rather than declining smoothly in density with increasing distance from the centre of the halo. Furthermore, the halo is almost certain to contain localised 'streams' of dark matter created when smaller dwarf galaxies are captured and tidally disrupted by the overwhelming gravitational influence of the Galaxy as a whole. One such stream is believed to be associated with the Sagittarius dwarf, a diffuse, low-mass galaxy which is located just on the far side of the galactic centre from where we are. The Sagittarius dwarf is actually orbiting over the poles of the Milky Way system, with long streams of stars preceding and trailing the main body, one of which (the leading tail)

This image shows one of the array of modules that comprise the DRIFT-II directional WIMP detector at Boulby Underground Laboratory. The inner detector is being loaded into a steel vacuum vessel, which will be filled with carbon disulphide target gas.

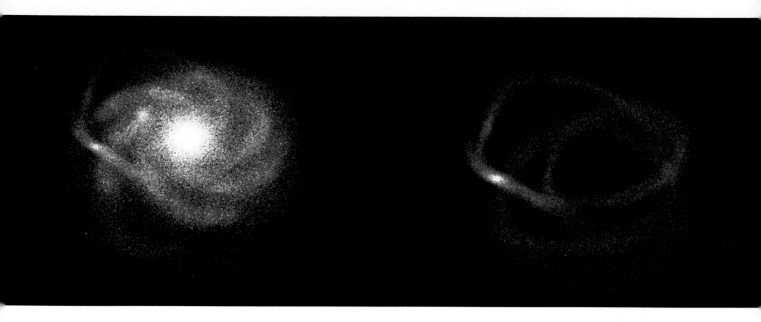

is heading towards the neighbourhood of the Sun. Kathcrine Freese, of the University of Michigan, and co-workers, have suggested that if the Sagittarius dwarf originally contained some dark matter, the leading tail should be showering dark matter down on the solar neighbourhood.

WIMP streams will superimpose additional peaks, in different directions, on top of the underlying average WIMP wind. Since the event rates for the halo and a stream will, in general, rise to a maximum at different times in the year, the contribution made by the stream will affect the peak date of the total WIMP signal. Freese and her colleagues have suggested that if, for example, the local density of the Sagittarius WIMP stream is around 4 percent of the total WIMP wind density, that would be enough to shift the peak of the annual modulation of the DAMA signal from 2 June to 25 May, which would actually produce better agreement with the DAMA data than a simple halo alone. By contrast, a 20 percent stream contribution would shift the peak date to around 30 March, in disagreement with the DAMA data. While other experiments which simply count number of events ought to be able to find evidence of the Sagittarius stream, and other such streams, the best prospect of directly detecting and identifying WIMP streams must surely lie with directional detec-

tors such as DRIFT and NEWAGE.

In the final analysis, directional detectors offer the best hope for conclusively demonstrating the presence (or absence) of the galactic WIMP halo, through which the Solar System is presumed to be travelling.

Looking for WIMP by-products

Instead of trying to detect WIMPs directly, some researchers are concentrating on looking for the by-products that are likely to be produced when WIMPs collide with each other. Since no one yet knows the precise nature of WIMPs (or even, for certain, whether or not they exist), particle physicists and cosmologists do not know whether each WIMP has an anti-WIMP opposite number, or if – as Supersymmetry predicts in the case of the neutralino – WIMPs are their own antiparticles. For significant amounts of annihilation to take place, WIMPs either must be their own antiparticles – so that annihilation takes place when two WIMPs collide with each other – or the present-day universe must contain approximately equal numbers of WIMPs and anti-WIMPs. Either way, the outcome of a collision will be the annihilation of both particles and the release of energetic photons (gamma-rays), or the production of short-lived particles which themselves decay into gamma-ray photons, energetic neutrinos, or other kinds

The Milky Way Galaxy is tearing streams of material from its nearest neighbour, the Canis Major dwarf galaxy (the compact bright patch towards the left of each image).
(Left) The location of the Canis Major dwarf in relation to the Sun (the bright patch above and left of centre) and the Milky Way Galaxy.
(Right) Simulations show that over a period of two billion years, the stream of stars lost from the Canis Major dwarf galaxy is able to wrap round our Galaxy three times.

of stable long-lived particles such as protons, antiprotons, electrons or positrons (anti-electrons). Whereas protons and electrons produced in this way would merge into the general population of matter, particles in the universe, photons, neutrinos, antiprotons and positrons are more likely to stand out above the general background. Indirect searches concentrate on looking for these by-products.

Because individual WIMP annihilations are likely to be extremely rare events, the best hope of detecting their by-products is to look towards places where the density of WIMPs is likely to be exceptionally high. If the orbit of a WIMP in a galactic halo passes through a celestial body, there is a very small but finite chance of its colliding with and bouncing off an atomic nucleus in that body. If the velocity of the WIMP after a collision is less than it was before, it may become trapped by that body's gravitational field and will eventually settle towards its centre. This process should lead to accumulations of WIMPs at, for example, the centre of the Earth, in the core of the Sun, or at the centres of galaxies (including the centre of our own Galaxy). The accumulated WIMPs will annihilate in pairs to yield a tell-tale signature of gamma-rays, neutrinos or antimat-ter. Detecting these signals will not be easy – the expected flux of gamma-rays from WIMP annihilations in the galactic centre has been estimated to be equivalent to the apparent brightness of a candle located on the planet Neptune!

Searching for antiparticles

Antiprotons and positrons produced by the annihilation or decay of dark matter particles will have a spread of energies (an energy spectrum) that extends up to a maximum value which is equivalent to the mass of the annihilating particles (the energy that would be released if the dark matter particle were converted completely to energy in accord-ance with Einstein's relationship, $E = mc^2$). Various experiments designed to detect cosmic antiprotons and positrons have, over the years, been flown on high-altitude balloon missions such as BESS (Balloon-borne Experiment

with a Superconducting Spectrometer) and HEAT (High Energy Antimatter Telescope). BESS has recently reported a possible, though rather marginal, excess of antiprotons at energies below about 1 GeV which could, just about, be due to the products of dark matter annihilations; whereas data acquired by the HEAT experiment show hints of an excess of positrons, at energies greater than 8 GeV, compared to the expected positron background. While these results may turn out to be completely explicable by nuclear reactions in interstellar space, they certainly do not rule out a possible contribution from the annihilation products of dark particles such as neutralinos or Kaluza–Klein particles.

Another way of tracing positrons is to measure the telltale gamma-ray signal, with an energy of 511 keV that results from the mutual annihilation of electrons and positrons. Electron–positron annihilation radiation from the galactic centre was first detected in the early 1970s and has been studied repeatedly since then, most recently, and most precisely, by the INTEGRAL (INTErnational Gamma-Ray Astrophysics Laboratory) satellite, which was launched by the European Space Agency in 2002. The data show a strong 511 keV signal coming from the region of the galactic centre which is, of course, where the greatest concentration of dark matter, and hence of dark matter annihilations, is to be expected. Once again, it may be that conventional proc-esses can wholly account for this signal, but a contribution from dark matter annihilations is certainly a possibility.

But while the observational evidence from antiproton and positron detections and from gamma-rays emitted by electron–positron anni-hilation offers some tantalising hints of WIMP annihilation, for the time being, the evidence remains decidedly tenuous.

WIMP-induced neutrinos

Muon neutrinos produced by the decay of WIMP annihilation products such as tau leptons, quarks, W and Z bosons and Higgs bosons are expected to have energies ranging from a few GeV (equivalent to several times

the mass of a proton) to a few TeV (equivalent
to several thousand times the mass of a
proton) – very much higher than the energies
of solar neutrinos. Rather than trying to detect
these neutrinos directly, some experimental
groups have concentrated on searching for
the by-products of interactions between high-
energy neutrinos and terrestrial rock. Muon
neutrinos are particularly interesting in this
respect because their interactions with rock
produce muons, which can travel considerable
distances through rock (by way of contrast,
electrons produced by electron neutrinos can
travel only very short distances through rock).
High energy muons produced in this way
will usually follow tracks that are reasonably
closely aligned with the paths followed by the
muon neutrinos that created them in the first
place, and will also carry much of the original
neutrino's energy.

Experimenters can therefore search for
evidence of WIMP annihilations by looking
for an excess of neutrino-induced muons
coming from the direction of the Earth's
centre, the Sun or the galactic centre, over the
background of muons created by atmospheric
neutrinos (neutrinos produced in the Earth's
atmosphere by cosmic rays – see Chapter 5).
To ensure that they are looking only at muons
created by neutrino interactions, experimenters use the Earth itself as a shield against
cosmic rays: they concentrate on measuring
only those muons that are travelling upwards
through the detector, having been created by

This artist's view shows
the INTEGRAL gamma-ray
satellite, which was launched
in October 2002.

neutrino interactions in the rock beneath the
level at which the detector is located. The task
facing experimenters is to separate out muons
produced by muon neutrinos that were created by WIMP annihilations inside the Earth,
the Sun or the core of the Galaxy, from muons
produced by interactions between atmospheric
neutrinos and terrestrial rock.

So far, none of the current generation of
neutrino detectors has found any statistically
significant evidence of excess neutrinos coming from the Earth, the Sun or the galactic
centre but they have, at least, been able to
set limits on the numbers of WIMP-annihilation neutrinos that are reaching the Earth.
However, over the next few years, experiments

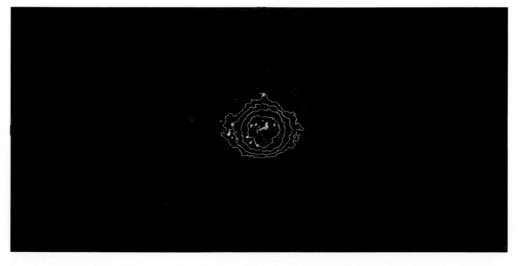

The contours on this all-sky
map show the distribution
of gamma-ray emission
resulting from electron-
positron annihilation, as
measured by the INTEGRAL
satellite. The emission is
concentrated in the direction
of the centre of our Galaxy.
One possibility (out of
several) is that the positrons
may originate from the
decay or annihilation of dark
matter particles.

– an array of 750 downward-looking detectors, buried at depths of up to 2500 metres in the Antarctic ice cap, which look for flashes of Čerenkov light in the surrounding ice. IceCube is a vastly bigger successor to AMANDA, which eventually will consist of 4200 detectors distributed within a cubic volume of ice measuring one kilometre along each face. Construction began in 2005 and is expected to take about six years to complete. Experiments such as these should go a long way towards establishing whether or not WIMP-annihilation does take place in the core of the Earth, in the Sun or in the galactic nucleus. Even if the results do not prove that WIMPs exist, at the very least they will tell us a great deal more about what WIMPs are not.

This artist's impression shows the strings of detectors that comprise the underwater 0.1 square kilometre ANTARES water Čerenkov telescope on the floor of the Mediterranean Sea.

This artist's conception shows the four 12-metre diameter optical reflectors of the VERITAS gamma-ray atmospheric Čerenkov array at Kitt Peak, Arizona.

such as ANTARES, AMANDA II and IceCube, are expected to achieve a hundredfold improvement in sensitivity to WIMP-annihilation neutrinos.

ANTARES is a huge neutrino telescope which is being constructed at a depth of about 2500 metres in the Mediterranean Sea near Toulouse, France. It will use downward-looking photomultipliers to pick up the flashes of Čerenkov light that are emitted by upward-moving muons created by neutrino interactions in seawater or in the rock below. AMANDA II is the current version of the Antarctic Muon and Neutrino Detector Array

Clues from gamma-rays

WIMP annihilation can produce gamma-rays in several different ways. First, the decay of short-lived particles which are themselves by-products of WIMP annihilations should radiate a continuous spectrum (a broad range of gamma-ray energies and wavelengths). The resulting spectrum of gamma-ray energies can extend up to, but not beyond, an energy value that is equivalent to the mass of the decaying WIMP. Consequently, the energy at which the gamma-ray spectrum 'cuts off' should give a strong indication of the mass of the annihilating WIMPs. Secondly, in situations where pairs of WIMPs collide and annihilate directly into gamma-ray photons, the resulting emission will be concentrated at one or more particular wavelengths and energies, so producing well-defined gamma-ray spectral lines.

Searches for WIMP-annihilation-induced gamma-rays have been undertaken from the ground and from space. Ground-based detection experiments use Atmospheric Čerenkov Telescopes (ACTs) which detect the faint blue Čerenkov light emitted by the shower of energetic particles that is sprayed off when an incoming high-energy gamma-ray interacts with the top of the atmosphere. Gamma-ray photons interact with atomic nuclei to create pairs of electrons and positrons which, through further collisions, create yet more

electrons and positrons as they descend lower in the atmosphere, so creating a cascade of particles which is known as an air shower. Since the charged particles in the air shower are highly energetic, they travel close to the speed of light in a vacuum and faster than the speed of light in air. Consequently (see Chapter 5) they emit photons that spread out in a cone along the direction in which the shower particles are heading.

An ACT will detect that light if it happens to lie within the so-called 'light pool' – the region where the cone of Čerenkov light meets the surface of the Earth, which typically is about 250 metres in diameter. The flashes of light last for only a few billionths of a second, so experimenters need to use exceedingly short exposures to image them. The resulting image shows the track of the air shower, which points back towards the source of the incoming gamma-ray that generated the shower. The direction of the gamma-ray source can be determined more precisely if several separate telescopes succeed in viewing the same shower from different points within the light pool. This Čerenkov imaging technique, which was pioneered at the Whipple Observatory in Arizona, in effect uses the atmosphere itself as a giant detector.

Examples of atmospheric Čerenkov telescopes, with marvellous acronyms, include CANGAROO, VERITAS, HESS and MAGIC. CANGAROO (Collaboration between Australia and Nippon for a GAmma-Ray Observatory in the Outback) is a joint Australian-Japanese project based at Woomera, Australia, which in its current version, consists of four 10-metre telescopes. VERITAS (Very Energetic Radiation Imaging Telescope Array System) is a new array of four 12-metre telescopes currently nearing completion at Kitt Peak, Arizona, and HESS (High Energy Stereoscopic System) is a similar array located in Namibia. MAGIC (Major Atmospheric Gamma Imaging Čerenkov telescope) is a 17-metre gamma-ray dish located on a mountain site in La Palma, Canary Isles. ACTs like these are typically sensitive to gamma-rays with energies ranging from 100 GeV to 10 TeV, and can very efficiently (usually better

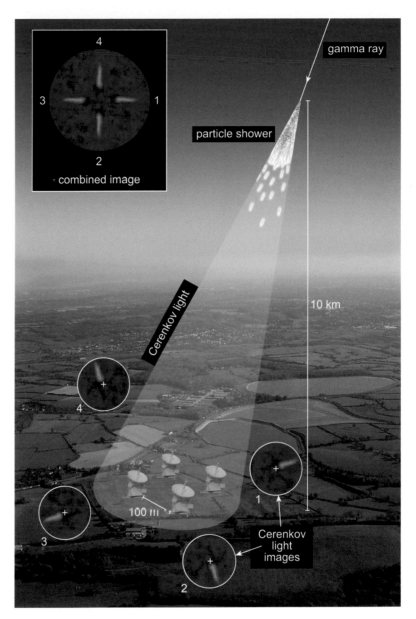

than 99 percent) distinguish between atmospheric showers produced by gamma-rays and showers produced by cosmic rays (which make up the dominant background).

Of the satellite-based detectors, the most significant so far have been EGRET, an instrument carried aboard the Compton Gamma-Ray Observatory which operated in orbit for nine years, from 1991 to 2000, observing gamma-rays in the 20 MeV to 30 GeV range, and INTEGRAL, which was launched in 2002. The next big mission is GLAST, a spacecraft funded by NASA, the US Department of Defense, together with France, Germany, Japan, Italy

Čerenkov light emitted by the shower of particles created when an energetic gamma-ray ploughs into the atmosphere, spreads out in a cone and briefly illuminates a small patch of ground. When the images obtained by telescopes that lie within this 'light pool' are combined (upper left), they reveal the direction from which the gamma-ray arrived.

This composite image shows the single 17-metre dish of the MAGIC atmospheric Čerenkov telescope for gamma-ray astronomy on La Palma, in the Canary Isles, seen at dusk. The largest instrument of its kind, its reflecting surface is made up of hundreds of individual sections.

and Sweden, which is scheduled for launching in August 2007. GLAST will observe gamma-rays in the 10 MeV to 100 GeV range with a field of view twice as great as, and a sensitivity at least 50 times better than, EGRET. However, the relatively small size of space-borne detectors limits the effectiveness of satellite experiments, especially at higher energies, where the flux of gamma-rays becomes progressively weaker.

Tantalising hints

In 2004, three of the Atmospheric Čerenkov Telescope groups – VERITAS, CANGAROO II and HESS – reported detections of very high energy gamma-rays coming from the direction of the galactic centre, results which could possibly be explained by WIMP annihilations but could equally well be accounted for by more 'ordinary' astrophysics. The gamma-ray spectrum recorded by CANGAROO matches reasonably well with what would be expected from the annihilation of WIMPs with masses in the region of 1–3 TeV/c² (about 1000–3000 times the mass of a proton), but the VERITAS and HESS results would require considerably heavier WIMPs (because the HESS observations show no obvious sign of a cut-off at the high energy end of the gamma-ray spectrum, WIMP masses of at least 12 TeV would be needed to account for the observations). Because the annihilation rates would have to be extremely high if any of these results were to be a con-

sequence of WIMP annihilation in the galactic centre, the jury is still very much out on the nature of the source of these excess signals.

Equally intriguing is the recent apparent detection by CACTUS (an Atmospheric Čerenkov Telescope based in California) of gamma-ray emission from the nearby Draco dwarf galaxy – a satellite of the Milky Way which appears to contain a very high proportion of dark matter. It seems to be just about within the bounds of possibility that the reported gamma-ray signal could be produced by dark matter annihilation, but other explanations are possible, and further observations at higher resolution will be needed before that possibility can be tested in detail.

The EGRET experiment, which mapped the distribution of gamma-rays across the whole sky, has also thrown up some tantalising data. As early as 1997, experimenters noticed that there seemed to be stronger diffuse (spread out rather than localised) gamma-ray emission at energies greater than 1 GeV than conventional models of the Galaxy had predicted. The data were re-analysed in 2004 by Wim de Boer of the University of Karlsruhe, Germany, with some intriguing results. De Boer suggests that, at energies greater than 0.5 GeV, an additional contribution to the gamma-ray flux by dark matter annihilation is needed to account for the excess emission in the gamma-ray spectrum at energies greater than 0.5 GeV. He estimates that the annihilating WIMPs would need to have masses in the region of 50–100 GeV/c² (the sort of mass range which is widely discussed in WIMP literature), and cross-sections in the region of 10^{-7} pb, values which lie within the grasp of current and forthcoming direct detection experiments.

The EGRET diffuse gamma-ray excess shows all the features that would be expected from WIMP annihilation in the dark matter halo. The signal is present in all sky directions and has the same spectrum in all directions. The intensity of the excess radiation falls off with the square of distance outwards from the galactic centre, which fits in with the inverse square distribution of dark matter density that is needed to account for the flat rotation curve of the galactic disc (the density of ordinary luminous matter falls off much more rapidly).

Intriguingly, the measurements also reveal two zones of strongly enhanced gamma-ray emission, at distances of 4 and 14 kiloparsecs (13,000 and 46,000 light-years) from the galactic centre. The 14 kiloparsec zone coincides with a ring of stars, discovered in 2003, which is believed to have been formed when a dwarf galaxy was torn apart by the gravitational influence of the Milky Way Galaxy. If the dwarf galaxy had originally contained a significant quantity of dark matter, its disruption would have enhanced the density of galactic dark matter at the location of the ring of stars. The position and shape of the 4 kiloparsec ring coincides with a known ring of molecular hydrogen gas which could have been pulled together by the gravitational influence of an underlying ring of dark matter.

De Boer and his colleagues calculated the amount of WIMP dark matter that would be needed to account for the excess gamma radiation observed in the rings, and found that it was just about right to reproduce a hitherto unexplained anomalous change in the slope of the galactic rotation curve at a distance of 11 kiloparsecs from the galactic centre. In the absence of dark matter, there would be far too little luminous mass to make so great an impact on the rotation curve.

While it would be premature to suggest that the EGRET results show definite evidence of galactic WIMPs, de Boer argues that they do provide 'an intriguing hint that dark matter is not so dark, but visible by its annihilation'.[1]

A grainy halo?
The possible existence of dark matter streams such as the Sagittarius stream, and toroidal rings such as those discussed by de Boer, imply that the detailed structure of the dark matter halo may be a great deal more complex than the standard halo model suggests. To add to the possible complexity, a supercomputer simulation of the evolution of dark matter structures carried out at the University of Zurich by Jurg Diemand, Ben Moore and J Stadel, and published in January 2005, indicates that the galactic dark matter halo may contain many trillions of small-scale dark matter blobs, or minihaloes. Assuming that the dark matter is composed primarily of neutralinos with masses of around 100 GeV/c^2, the Zurich group found that the first structures began to form at a redshift of 60 (when the universe was still less than 2 percent of its present size), and that these were dark matter haloes with masses in the region of one millionth of a solar mass (broadly comparable to the mass of the Earth) and diameters comparable to the size of the present-day Solar System. They found that, at distances from the galactic centre greater than about 10,000 light-years, these minihaloes could survive intact right through to the present day, and that there should be about 10^{15} (a thousand trillion) of them embedded within the overall galactic halo.

If they are right, a significant fraction of the dark matter mass in our region of the Galaxy may be confined within these kinds of objects. The Zurich group predict that, on average, the Earth will encounter a dark matter minihalo about once every 10,000 years and will take about 50 years to pass through it. The corollary is that the Earth will spend most of its time in relatively low-density regions of the overall halo; this would affect the predicted WIMP flux in the overall WIMP wind. A further point made by Diemand, Moore and Stadel is that that the nearest minihaloes will be among the brightest sources of gamma-rays from WIMP annihilations. Although minihaloes may be

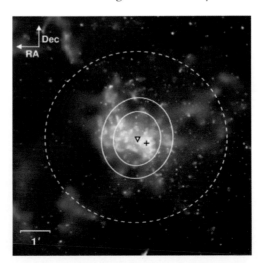

The solid contours, superimposed on a Chandra x-ray image of the centre of our Galaxy, show the location of a strong, very high-energy, gamma-ray source detected by the HESS atmospheric Čerenkov telescopes.

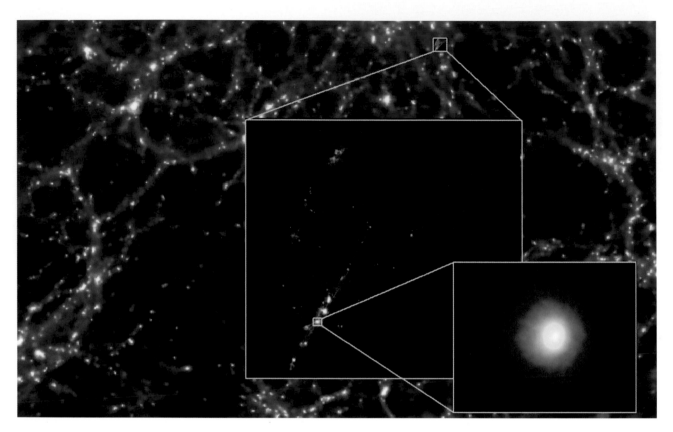

This image shows a zoom into the first object to form in the universe, according to a computation by a University of Zurich team. The two inset regions are successive enlargements of a tiny region within the 10,000 light-year diameter blue region. The inset on the lower right shows a single Earth-mass dark matter structure the size of the Solar System.

detectable via gravitational effects such as lensing or dynamic perturbations (for example the cumulative effect of encounters of this kind may disturb the Oort cloud from which long-period comets emerge), the best prospect for detecting them lies in looking for atmospheric Čerenkov radiation from gamma-rays generated by WIMP annihilation in the cores of these bizarre objects.

Current status of the WIMP hunt

The search for WIMPs is proceeding on a wide range of fronts. Attempts to detect them directly, by identifying WIMP-induced recoils of atomic nuclei in heavily shielded subterranean detectors have so far failed to make any positive identifications, though the sensitivities of these experiments are improving year by year. The claim by the DAMA team to have detected the expected annual variation in WIMP-induced nuclear recoils remains controversial but, despite being contradicted by other experiments, shows no sign of going away. Indirect detection experiments, which look for the by-products of WIMP annihilation processes, have

yet to produce any convincing evidence of the existence of WIMPs, although some tantalising hints seem to be emerging from antiparticle and gamma-ray observations. The range and ingenuity of the experiments is breathtaking. From within subterranean caverns, from the frozen depths of Antarctica, from the surface of our planet and from space, WIMP hunters are using as detectors everything from purpose-built laboratory-based and satellite-borne instruments to the oceans, the Antarctic ice cap, the atmosphere and the rocks of the Earth itself, in their drive to track down and identify this most elusive form of dark matter. If WIMPs exist, surely they deserve to find them.

CHAPTER EIGHT

Matter is not Enough

Whether luminous or dark, baryonic or non-baryonic, 'hot' or 'cold', all forms of matter exert a gravitational attraction. As we saw in Chapter 2, if matter and radiation (which also has a gravitational influence) are the sole constituents of the universe, its long-term future and ultimate fate will depend on whether its overall mean density is greater than, less than, or equal to the so-called critical density. If its mean density is greater than critical, the universe will be 'closed'; gravity will eventually slow the expansion of the universe to a halt, and the universe thereafter will collapse into a final 'Big Crunch'. If its mean density is less than critical, the universe will be 'open'; the rate of expansion will slow down towards a steady value, and the universe will continue to expand forever. If, however, the mean density is precisely equal to the critical density, the universe will be 'flat'; sitting on the fence between the open and closed options, it will just, but only just, be able to expand forever. As Einstein showed, the overall caaurvature, or geometry, of space depends on the density and distribution of matter and energy. In a closed universe space is positively-curved, analogous to the surface of a balloon, whereas in an open universe space is negatively-curved, analogous to the shape of a saddle. In a 'flat', or 'Einstein–de Sitter', universe, the overall net curvature is zero so that the large-scale geometry of space is analogous to that of a flat tabletop, and the familiar rules of Euclidean geometry apply. The destiny of a universe filled with matter and radiation alone is intimately entwined with its geometry.

As we also saw in Chapter 2, cosmologists use the symbol Ω (Omega) to denote the ratio of the actual mean density to the critical density, and call this ratio the *density parameter*. If Ω is greater than one ($\Omega > 1$), the mean density is greater than the critical density and the universe is closed, whereas if Ω is less than one ($\Omega < 1$), the mean density is less than critical and the universe is open. But if Ω is equal to one ($\Omega = 1$), the mean density is precisely equal to the critical density, and the universe is flat. By the 1970s, studies of the rotation curves of galaxies and the dynamics of galaxy clusters had shown that Ω is almost certainly greater than 0.1. At the other end of the range, it was equally evident that Ω could not be greater than 10 (corresponding to a density 10 times greater than critical), for if it were, the universe would already have collapsed into a Big Crunch, and we would not be here to debate the issue. Whatever the precise value of Omega may be, cosmologists were intrigued and puzzled by the fact that it appears to be remarkably close to one.

Why should this be in any way perplexing? The argument runs as follows. If the universe started out with a value of Omega precisely equal to one (so that the mean density in the earliest instants of its history was precisely equal to the value of the critical density at that time), then the value of Omega will always be exactly equal to one. As the universe expands, and matter and radiation become more dilute, the actual mean density and the calculated value of the critical density will decrease at exactly the same rate, so that the ratio of the two will remain unchanged and Omega, therefore, will always be equal to one. But if the initial value of Omega had been minutely greater or less than one, its value would diverge ever further from one as the expansion continued, and by the present day would differ from one by an enormous factor.

By way of analogy, we can think about the concept of escape velocity. If we throw a ball upwards, it will rise to a certain height, and then fall back to Earth. If we throw it faster, it will rise further before gravity wins the battle and pulls it back. If we could throw it fast enough, it would continue to recede forever; although it would slow down, it would never come to a halt and would never fall back to Earth. Escape velocity is the minimum speed at which a projectile must be fired in order to continue to recede forever, and not fall back. The value of escape velocity at any particular point depends on the mass of the body from which the projectile is trying to 'escape' and on the projectile's distance from the centre of that body – the greater the distance, the lower the escape velocity. At the surface of the Earth, the escape velocity is 11.2 kilometres per second (about 40,000 kilometres per hour). If we launch a spacecraft at that speed (ignoring the effects of air resistance) and let it coast through space thereafter, gravity will slow it down but, because the value of escape velocity decreases with increasing distance, its speed will always be equal to the local value of escape velocity. Its speed will decline ever closer to zero, but will not become zero until the spacecraft has receded to an infinite distance. If the launch speed is less than escape velocity, the spacecraft will rise to a maximum height and then fall back. But if the spacecraft is launched with an initial speed greater than escape velocity, its speed of recession will decline towards a constant value (rather than zero) and it will continue to recede at a finite speed forever.

If the projectile's initial speed is exactly equal to escape velocity, it will always be exactly equal to the local value of escape velocity, and the ratio of its actual velocity to the escape velocity will always be exactly equal to one. On the other hand, if its initial speed is slightly greater than escape velocity the initial ratio of these two speeds will be slightly greater than one. As the spacecraft recedes to ever greater distances, its speed will approach ever closer to a constant value, and the local value of escape velocity will continue to decrease.

Therefore, the ratio of actual velocity to escape velocity will become larger and larger, eventually becoming enormous. Conversely, if the initial speed is slightly less than escape velocity, so that the ratio of actual velocity to escape velocity is slightly less than one, the ratio will become smaller and smaller as the receding spacecraft slows down. At some particular distance (where the local escape velocity still has a finite value), its speed of recession will drop to zero, and the ratio itself will become zero (even if the initial ratio had been, say, 0.9999999, it will eventually drop to zero).

Therefore, if we observe a very distant spacecraft that is receding at the local escape velocity, we know that it must have started out with a velocity that was exactly equal to escape velocity.

Flatness, fine tuning and the horizon
The universe behaves in the same sort of way. We can think of a flat universe as one which has a mean density exactly equal to the critical density, and which is expanding at its own 'escape velocity'. Similarly, a closed universe has a density greater than critical and is expanding slower than escape velocity (so that it will eventually collapse on itself), whereas an open universe has a density less than critical, and is expanding faster than its own escape velocity. Unless the universe started off expanding at, or extraordinarily close to, its own escape velocity, it cannot be expanding at, or close to, its escape velocity now. If the mean density of the present-day universe differs from the critical density by a factor of less than 10 ($0.1 < \Omega < 10$), the universe must have started out with a density that was incredibly close to the critical density. Yet there is nothing in the standard Big Bang model that requires the universe to have any specific value of initial density, let alone a density that matches so closely with the critical value. This conundrum, which is known as the fine-tuning problem, or flatness problem (because, if the density of the universe is so close to critical, then the curvature of space must be very close to being flat), has prompted many theoreticians to argue that the actual value of Omega must be precisely

equal to, or indistinguishably close to, one.

Apart from localised temperature variations of a few parts in 100,000, the microwave background radiation is remarkably uniform and isotropic – it looks the same in every direction. To within a few parts in 100,000, opposite regions of the sky have exactly the same brightness and temperature. This raises a further issue, which is known as the *horizon problem*. At any time during the history of the universe, the maximum distance to which a particle (or observer) can 'see', or across which any cause-and-effect influence can have propagated (assuming that no signal or influence can travel faster than light), is called the *horizon distance*. It is equal to the age of the universe multiplied by the speed of light. When the universe was one second old, the horizon distance was one light-second (300,000 kilometres) whereas, 380,000 years later, when space became transparent and the microwave background radiation was released, it had expanded to 380,000 light-years. Today, if the universe is 13.7 billion years old, the horizon distance is 13.7 billion light-years. If galaxies exist beyond a range of 13.7 billion light-years, we cannot see them, because there has not been enough time in the history of the universe for their light to reach us. Likewise, when the universe was 380,000 years old, no particle or localised region of space could have been aware of the existence of any other particles or regions of space beyond a range of 380,000 light-years, nor could they have been influenced by matter or radiation beyond that range.

At the time when the microwave background radiation was released, the volume of space that constitutes the presently observable universe would have had about one thousandth of its present radius – roughly 10 million light-years – which was very much larger than the 380,000 light-year horizon distance. Therefore, at that time, it would have consisted of a large number of regions that were completely unaware of each other. The microwave sky that we see today consists, in effect, of a mosaic of patches which preserves a fossil record of the physical conditions that existed, 380,000 years after the beginning of time, in each of these

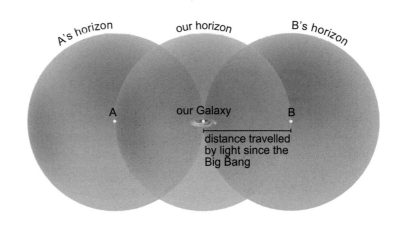

apparently unconnected regions of space. If they had no way of communicating with each other and ironing out any initial differences, how could all the distinct regions of space that make up the microwave 'sky' possess exactly the same properties? Even now, more than 13 billion years later, those parts of the microwave background that lie on opposite sides of our sky should not have had time to 'see' each other or influence each other. Either the universe, by sheer chance, just happened to start out smooth, uniform and identical everywhere, or there must have been some physical process that brought about this state of affairs.

Inflation offers an answer

In the early 1980s, in an attempt to get round these problems and some related issues in particle physics, American physicist Alan Guth introduced the concept of the *inflationary universe*. According to the inflationary hypothesis, for a brief period in its very early history, the universe experienced a dramatic phase of accelerating expansion during which all distances in the universe expanded, or 'inflated', by a colossal factor. At the end of the inflationary era, the universe reverted to a gently decelerating rate of expansion, governed by the effects of matter and radiation. While inflation was taking place, the expansion would have been 'exponential' – with all distances increasing by the same factor in each successive interval of time. If the 'doubling time' had

The radius of the observable horizon is equal to the maximum distance that light can have travelled since the Big Bang. Whereas we can see points A and B, which are on opposite sides of our horizon, an observer at point A cannot yet see point B because there has not been enough time for light to travel from A to B.

been, say, 10^{-34} seconds, then all distances (and separations between particles) would have doubled in 10^{-34} seconds, increased by a factor of four in 2×10^{-34} seconds, grown by a factor of eight in 3×10^{-34} seconds, and so on. After a hundred doublings (which would have taken a total of 10^{-32} seconds), all distances would have expanded by a factor of around 10^{30} (one million trillion trillion). So dramatic a degree of accelerated expansion is hard to visualise. It is roughly equivalent to the full stop at the end of this sentence expanding to become larger than the entire presently observable universe in considerably less time than it takes to blink an eye.

The process of inflation is believed to have begun about 10^{-35} seconds after the beginning of time and to have lasted, perhaps, for around 10^{-32} seconds. In order to produce a universe which seems smooth, uniform and isotropic today, the original microscopic volume of space, which corresponds now to the currently observable universe, would have had to inflate by a factor of at least 10^{27}. Had that been the case, the original volume, which 10^{-35} seconds after the beginning of time would have had a radius of about 10^{-27} metres, would have grown after 10^{-32} seconds to around a metre in radius. Thereafter, the ongoing expansion of space would have stretched that uniform region of space to a size at least as large as the presently observable universe. However, the inflation factor could easily have been much greater, 10^{50}, 10^{100}, or more, in which case the observ-

(a) Whereas the observable horizon round a particular point grows steadily (at the speed of light), inflation causes a tiny uniform region of space which initially (1) was smaller than the horizon to swell so dramatically that it becomes, and remains today (5) far larger than the observable horizon.
(b) A graphical representation of the expansion factor during inflation.
(c) Inflation stretches out the curvature of space so greatly that it becomes indistinguishable from 'flat'.

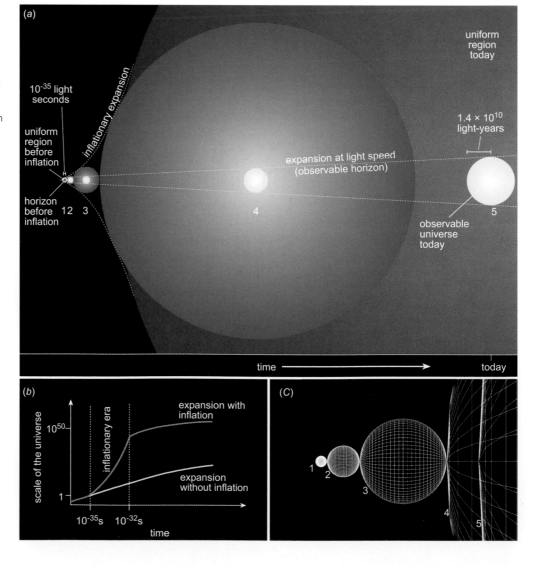

able universe which we see today is merely a microscopic portion of a very much larger inflated volume of essentially uniform space. Little wonder, then, that the universe looks the same in every direction.

By postulating that the observable universe formed from a region of space that, prior to inflation, had ironed out differences between its constituent parts, the inflationary hypothesis neatly solves the horizon problem. It also takes care of the flatness (or fine tuning) problem. The inflation of space by such a large factor would have flattened out its curvature so severely as to make it indistinguishable from 'flat'. When we look at the observable universe, we are looking at a tiny portion of the whole. Just as when we measure a small piece of the Earth's surface (say a circle 1 metre in radius) we conclude that the Earth is flat; so we cannot detect any net curvature in the space embraced by the limits of the observable universe. If inflation drove the curvature of space inexorably to being indistinguishably close to flat, so it would have driven the mean density to exactly, or indistinguishably close to, the critical density. In the inflationary universe, Omega should have a value indistinguishably close to one.

What could have caused the early universe to behave in this way? The answer may lie in intimate links that are emerging between high-energy particle physics and the high-temperature phase of the early instants of the Big Bang. High temperatures correspond to high particle energies, and in those earliest instants, particle energies were many orders of magnitude higher than anything that can be produced in a terrestrial particle accelerator. In the present low-energy state of the universe, the behaviour of matter and radiation is governed by the four forces of nature that were described in Chapter 1: (in order of decreasing strength) the strong nuclear force, the electromagnetic force, the weak nuclear force and gravitation. Theory suggests that at high enough energies, the differences between the forces begin to disappear. Experiments have already shown that at particle energies of around 100 GeV – energies comparable with the rest-mass

energies of the W and Z bosons described in Chapter 5 – the electromagnetic and weak nuclear forces lose their identities and behave as a single unified 'electroweak' force. Particle energies of that order correspond to a temperature of around 10^{15} K (one thousand trillion degrees), which is similar to the temperature of the universe at about one ten-billionth of a second after the beginning of time.

Grand Unified (GUT) theories predict that at even higher energies (10^{14}–10^{15} GeV) and temperatures (around 10^{27}–10^{28} K), which existed throughout the universe around 10^{-35} seconds after the beginning of time, the electroweak and strong forces will lose their separate identities and will merge into a single force. Although no particular theory has yet been confirmed (among the contenders are Supersymmetry and string theory – a theory in which particles are represented by different vibrational states of utterly microscopic little strings), there is wide agreement among physicists that current theories are heading in the right direction. The final unification, a 'theory of everything' (TOE) that embraces all four forces and the fundamental particles, has so far eluded theoreticians, largely because gravity is so much weaker than all the other forces and appears to behave in a rather different way. Whereas there are sound workable quantum theories of the strong, electromagnetic and weak forces (where forces are communicated by 'force-carrying' particles), gravity has so far resisted attempts to fit it into a quantum framework. Nevertheless, there is a general consensus that at particle energies of around 10^{19} GeV (corresponding to temperatures of around 10^{32} K), such as would have prevailed in the universe about 10^{-43} seconds after the initial event, gravity, too, will merge with the other forces.

A dramatic change in the state (or 'phase') of the universe would have occurred when the temperature dropped below the level at which the strong and electroweak forces were united. The splitting of the GUT force into the strong and electroweak forces has been likened to the everyday phase change that occurs when water vapour condenses to form liquid water, or when water freezes to become ice. Each of these changes releases heat energy (called latent heat). By analogy, the change from a high-energy state (unified forces) to a low-energy state (separate forces) may have provided the energy that drove the universe into a phase of exponential (accelerating) expansion that persisted until the phase transition was complete.[1] Thereafter, the universe reverted to expanding at a gradually decelerating rate, controlled by the influence of gravity.

In addition to providing neat solutions for the flatness and horizon problems, the inflationary hypothesis has great potential for resolving other issues, too – in particular, the origin of structure (galaxies, clusters, filament, walls and voids) in the universe. During the era of inflation, tiny localised variations (called *quantum fluctuations*) in the otherwise uniform inflating bubble would have been stretched so severely by the accelerating expansion that they would have become large enough to become the seeds from which galaxies, clusters and larger-scale structures evolved. Because fluctuations that formed early in the inflationary era would have blown up by a larger factor than those that formed later (the earliest fluctuations would have been stretched by the full extent of inflation whereas those that formed later would been stretched only by the degree of inflation that took place between the time of formation and the end of the inflationary era), inflation would have produced a range (or spectrum) of different-sized fluctuations. These inflated quantum fluctuations became imprinted on the cosmic microwave background in the form of localised patches of marginally enhanced, or reduced, temperature (called temperature fluctuations) which correspond to underlying fractional differences in the density of matter (density fluctuations) that eventually were pulled in on themselves by gravity to create protogalaxies, galaxies and clusters.

Nice theory – but what about the density?

Because of its success in overcoming the horizon and flatness problems and in naturally generating temperature and density fluctuations on the range of scales that is seen in the cosmic microwave background, many cosmologists and particle physicists became firmly wedded to the idea that inflation did indeed take place a microscopic second after the beginning of time. But if that is correct, then the large-scale curvature of space must be zero, or virtually zero. The universe should be geometrically flat, and the overall mean density of the matter, radiation, and any other forms of energy that it may contain, must collectively add up to the critical density, or a density which is exceedingly close to critical. Ω must be exactly equal to one, or have a value exceedingly close to one.

The stars and glowing gas clouds that we see in the optically visible parts of galaxies provide less than 1 percent of the critical density and baryons, in total, somewhere between 4 and 5 percent. For a long time, particularly during the 1980s and early 1990s, many theoreticians asserted, felt, or hoped, that dark matter would make up the shortfall – that there would be enough dark matter to ensure that the overall density was equal to the critical value. For that to be so, non-baryonic dark matter – most probably cold dark matter – would have to comprise 95 percent of the total. During the 1980s and early 1990s, the favoured paradigm was 'inflation plus cold dark matter': inflation provided the density fluctuations from which structure grew, and the dominant gravitational influence of cold dark matter determined the way in which matter aggregated together to build cosmic structures from the bottom up (small structures first, large ones later).

Yet, despite its successes, the model had its problems. Simulations of structure formation in a critical density cold dark matter universe, though much more successful in this respect than hot dark matter simulations, still predicted levels of matter clustering that did not seem faithfully to reproduce what the observers were seeing. Furthermore, if the density of the universe were indeed equal to the critical density, there appeared to be an embarrassing 'age problem': the age of the universe (the time taken to expand to its present size) seemed to be less than the age of its oldest constituent stars (the issue will be explored in Chapter 10). To make matters worse, as we have seen in previous chapters, there was a strong and growing body of evidence which indicated that, although dark matter exists in substantial quantities, the total density of matter in all its forms (luminous and dark, baryonic and non-baryonic) is less than 30 percent of the critical value.

Were the observers wrong, the theoreticians wrong, a mixture of both, or was there another ingredient to the content of the universe that needed to be taken into account? What was needed was a clear-cut measurement of the curvature of space.

Curvature and the microwave background

Vital clues turned out to be embedded within the cosmic microwave background (CMB) which, as we saw in Chapter 2, consists of photons from the Big Bang fireball which became free to spread out through the expanding volume of space at the time of 'recombination', when atomic nuclei captured electrons to form complete atoms. In so doing, atomic nuclei mopped up the all-pervading 'sea' of electrons which, prior to that time, had been responsible for scattering photons so frequently that they could travel very little distance at all between successive collisions. The presently observable microwave background consists of photons which were scattered for the last time at the epoch of recombination and which have since been redshifted to microwave wavelengths by the ongoing expansion of space. For that reason, the microwave background is also referred to as the 'surface of last scattering'. In effect, when we look out at the microwave background in any direction, we are looking outwards in space and nearly 14 billion years back in time towards this 'surface'. We cannot look further back towards the Big Bang because, before recombination, space was opaque to electromagnetic radiation. The surface of last scattering is like an impenetrable foggy shell, some 14 billion light-years in radius, which surrounds us. Imprinted on that

'surface' is information about the physical state of the universe as it was when the background photons last interacted with matter, some 380,000 years after the beginning of time.

Because of the Sun's motion around the centre of the Galaxy, and the motion of the Galaxy caused by the gravitational influence of neighbouring concentrations of matter, the Earth moves relative to the sea of microwave photons. Radiation arriving from the direction in which the Solar System is heading is fractionally blueshifted (squeezed to shorter wavelengths) and appears slightly hotter, whereas radiation arriving from the opposite direction is fractionally redshifted (stretched to longer wavelengths) and appears marginally cooler. The measured temperature difference between opposite sides of the sky, which is known as the CMB dipole, shows that the Solar System is moving relative to the rest of the observable universe at about 370 kilometres per second.

When the effect of the dipole is removed, the CMB turns out to be remarkably smooth and isotropic (it has equal temperature and intensity in every direction). But in order for galaxies and clusters to have formed, there must have existed, at the time of recombination, density fluctuations (localised regions of marginally enhanced density) which would eventually collapse under the action of gravity to form dark matter haloes, galaxies and clusters. These density fluctuations would have given rise to temperature fluctuations (patches of marginally enhanced and reduced temperature in the CMB). Following the discovery of the CMB, astronomers began to search for these temperature fluctuations, but by the late 1980s, by which time no temperature variations as large as one part in 10,000 had been found, theoreticians were beginning to become worried. If there were no temperature fluctuations and density fluctuations, how could galaxies have come into existence?

To the great relief of much of the astronomical community, temperature fluctuations were finally revealed on 24 April 1992 by an American satellite called COBE (COsmic Background Explorer), which had been launched in 1990. The localised temperature

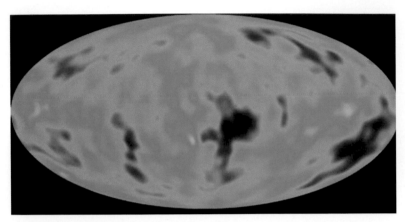

This all-sky map shows the marginally warmer and cooler patches (shown as colour variations) in the cosmic microwave background radiation which were detected for the first time, in 1992, by NASA's COBE mission.

differences that COBE found were very small – only about 30 microkelvins (30 μK), or about one part in a hundred thousand. This implied that the contrast in density between the associated density fluctuations in ordinary (baryonic) matter and their surroundings were too slight for the gravitational pull of baryonic matter alone to have caused them to collapse and form galaxies and clusters within the time that has elapsed since the Big Bang (see 'The growth of density fluctuations', facing page). This in itself pointed to the need for substantial amounts of non-baryonic dark matter which, unlike baryons, did not interact with photons, and therefore could clump together much more readily under the influence of gravity. Dark matter clumps provided the gravitational wells into which baryonic matter subsequently fell to create the protogalaxies from which larger galaxies and clusters were eventually assembled.

Although highly sensitive, COBE had poor angular resolution. Consequently, it could only reveal patches with angular sizes larger than 7 degrees across. Much finer resolution would be needed to reveal the far smaller seeds from which galaxies and clusters formed. From 1992 onwards, several ground-based and balloon-borne experiments have managed to resolve smaller scale features in the CMB, though admittedly only over relatively small localised areas on the microwave sky.

Temperature fluctuations of large angular size (more than a few degrees across) correspond to primordial density fluctuations which were larger than the horizon distance (the distance across which light could have travelled) at the time of recombination and the decoupling of matter and radiation. Created by the enormous stretching of space that was caused by inflation, they were so large that, at the time of decoupling, different regions within them were too far apart to 'see', interact with, or in any way be influenced by, each other. Consequently, their imprints on the CMB reveal large-scale fluctuations much as they were at the end of the inflationary era. The temperature differences that are associated with large-scale density fluctuations are caused by gravity; photons 'climbing out' of these regions lose energy and are redshifted towards longer wavelengths, so giving rise to cooler patches on the microwave sky. On these large scales, cooler patches correspond to denser regions and warmer regions to less dense regions.

Temperature fluctuations smaller than a degree or so in angular size correspond to clumps of matter and radiation which, at the time of decoupling, were *smaller* than the horizon distance. Consequently, these small-scale fluctuations *had been* affected by physical processes and interactions between their constituent particles and photons during the 380,000 years that elapsed between the end of inflation and the time of recombination; this is reflected in the imprint they made on the last scattering surface. In any part of the primeval fireball which, prior to decoupling, was denser than average (an 'overdense' region) and smaller than the horizon distance, gravity would cause particles of matter to fall together. But the pressure exerted by interactions between particles and photons would eventually halt the collapse and cause the matter to rebound then start to collapse again, thus setting up a series of inward and outward motions called acoustic oscillations. Just as the vibrations of a drum propagate through air at the speed of sound, so acoustic vibrations in the primeval mix of matter and radiation propagate at the speed of sound in that medium, which turns out to be extremely high – nearly 60 percent of the speed of light. The small-scale temperature fluctuations in the

The growth of density fluctuations

A density fluctuation is a region of space where the density is higher than average. If the mean density of the universe as a whole is equal to the critical density ($\Omega = 1$), then if a localised patch of space contains a greater than average density of matter, the density within that region will be greater than critical ($\Omega > 1$) and it will behave like a tiny 'closed universe' within the universe at large. Gravity will slow its rate of expansion relative to the expansion of the universe as a whole and, as a result, its density will decline more slowly, as time goes by, than the mean density of its surroundings; therefore the density contrast between this 'density fluctuation' and its surroundings will become progressively greater. Eventually gravity will halt its expansion and cause it to collapse on itself to create a self-contained object such as a protogalaxy.

In a scenario such as this, where density fluctuations grow under the influence of gravity, the density contrast grows in direct proportion to the scale factor (the expansion factor) of the universe; this is called 'linear growth'. In the time that it takes for the universe to expand by a factor of two, the density contrast in the initial fluctuation also grows by a factor of two; if the universe expands by a factor of 10, the density contrast grows by a factor of 10; and so on. Between the epoch of recombination and the present day, the scale of the universe has increased by a factor of just over 1000. However, if – as the measurements of the temperature fluctuations indicate – the density contrast in baryonic matter fluctuations at that time was only about one part in 100,000, then that contrast would only have grown to around one part in 100 by now – not enough to make protogalaxies, galaxies and clusters.

Because baryons and photons interacted so closely with each other in the Big Bang fireball, fluctuations in the density of baryonic matter could not begin to grow – under the action of gravity – until after that close coupling was broken. But cold dark matter particles did not interact with photons in this way (and only very rarely collided with each other or with baryons) and so could respond much more readily to gravitational forces. Consequently clumps of cold dark matter could begin to grow in density contrast, unimpeded by interactions with photons and baryons, much earlier than clumps of baryons could do. As a result, the density contrast between non-baryonic dark matter clumps and their surroundings at the time of recombination would have been substantially greater than the density contrasts for baryons, and the growth of structure could proceed much faster than it would have done had the universe contained baryons alone.

According to the cold dark matter scenario, the smaller the diameter of the density fluctuation, the greater the contrast between its density and the density of its surroundings. Consequently, smaller clumps collapse first to form low-mass objects which subsequently merge together to form successively more massive objects. Within this scenario, structures in the universe grow in a hierarchical fashion 'from the bottom up'.

microwave background represent a snapshot of the acoustic oscillations as they were at the time of recombination. Because the oscillating matter-radiation mix was hottest at greatest compression and coolest at maximum rarefaction, small-scale warm patches in the CMB correspond to regions which were at maximum compression, and cooler patches to regions that were at greatest rarefaction at the time when the decoupling of matter and radiation occurred.

The most prominent features in the CMB correspond to regions that were at maximum compression – halfway through their *first*

oscillation cycle – at the time of decoupling. From what we know of the density, pressure and temperature at that time, the physical size of these regions can be calculated (they are about the size of the sound horizon – the maximum distance that a sound wave could have travelled at the time of decoupling – which is just under 60 percent of the horizon distance for light at that time). The size of these regions provides a natural ruler. Since the time when the microwave background was released, nearly 14 billion years ago, the universe has expanded by a factor of more than a thousand (1089 according to WMAP), and a region which

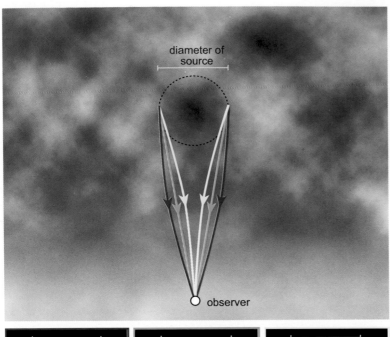

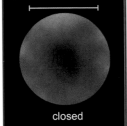 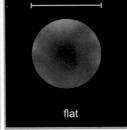 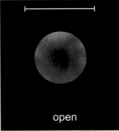

apparent diameter of source

The observed apparent size of a patch in the cosmic microwave background depends on whether space is positively-curved (closed), flat or negatively-curved (open).

corresponded to the diameter of the sound horizon at decoupling has expanded by the same proportion. When we observe the CMB today we are, in effect, looking out across a distance of around 14 billion light-years to the last scattering surface. Knowing the physical size of the 'ruler' (the size of the most prominent patches in the CMB) and its distance (nearly 14 billion light-years) we can work out its expected angular size. It turns out that if space is flat (so that rays of light travel in straight lines), the angular diameter of the most prominent temperature fluctuations should be about one degree (about twice the apparent size of the Moon in the sky).

Provided that instruments used to map the microwave background can resolve features smaller than a degree across, they can then compare the measured angular sizes of the warmer and cooler patches with

their theoretically predicted sizes. If space is flat, as inflationary theory requires, the most prominent patches should have an angular diameter of about a degree. But if space is positively curved, as in a closed universe, that type of curvature would act like a convex lens, or magnifying glass; the bending of light would make the patches look larger. On the other hand, if space were negatively-curved, as in an open universe, the curvature would act like a concave lens (a reducing lens) and make the patches look smaller.

By the late 1990s, ground-based microwave receivers had achieved angular resolutions much better than that of COBE, and in 1999, one particular experiment called MAT/TOCO (the Mobile Anisotropy Telescope, operating at an altitude of 5200 metres on Cerro Toco in the Chilean Andes), produced results that suggested the most prominent patches had angular sizes of about one degree. The clincher came with the publication, in the following year, of spectacular results that had been obtained by two balloon-borne experiments called BOOMERanG and MAXIMA. High altitude balloons are capable of carrying instruments to heights in the region of 35–40 kilometres and thereby getting above about 99 percent of the Earth's atmosphere. Launched from the McMurdo Research Station in Antarctica in December 1998, BOOMERanG succeeded in measuring about 2.5 percent of the whole microwave sky with a resolution of 0.2 degrees (and was therefore able to pick out details about 35 times smaller than anything that COBE had been able to reveal). MAXIMA made two flights of shorter duration, in August 1998 and June 1999, high above Texas, and mapped, with similar resolution, two different regions of sky with a total area equivalent to 1.5 percent of the entire sky.

Despite some detailed differences in their data, both sets of results agreed on one key issue – the angular size of the most prominent patches in the microwave background is just less than 1 degree – exactly as expected for a flat-space universe. This result has been confirmed by every subsequent ground-based, balloon-borne or space-based experiment. The

The BOOMERanG Telescope being prepared for launch in December 1998. With Mt. Erebus in the background, experimenters are inflating the 1 million cubic metre balloon which carried the telescope on its 10-day trip around the Antarctic continent.

conclusion seems very clear. The universe in which we live has a flat (or very nearly flat) geometry, and the value of Omega is equal to, or very close to, one. We live in a flat-space critical density universe, just as inflationary theory predicts.

Cosmic ripples tell a detailed story

High-resolution maps reveal temperature fluctuations with many different angular sizes. When cosmologists plot the relative brightness of these fluctuations against angular size they obtain a graph, called a *power spectrum*, which displays a series of peaks and troughs. The angular size of the first, and most prominent, peak in the temperature power spectrum – which corresponds to the size of the most prominent patches on the sky – is determined primarily by the overall geometry of the universe and, as we have already seen, seems to indicate that we live in a flat universe. Like overtones produced in a musical instrument, smaller peaks are expected to occur at about a half and a third, and so on, of the angular scale of the largest peak. The first peak, as we have seen, corresponds to an oscillation in the matter-radiation mix that has fallen inwards to maximum compression for the first time. The second corresponds to matter that has rebounded to minimum compression, the third to matter that has fallen inwards for the second time, and so on. Thus, odd-numbered peaks (first, third, and so on) correspond to compression and even-numbered peaks (second, fourth, and so on) to rarefaction.

The relative brightnesses, temperatures, sizes and distribution of the warmer and cooler patches in the microwave background depend

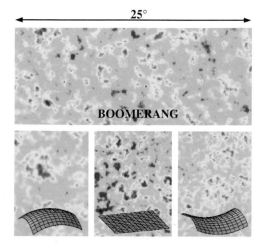

BOOMERanG images determine the geometry of space. From left to right, the three panels along the bottom show the sizes of the dominant warm and cool spots in the microwave background that would be expected if the universe were closed (positively-curved), flat, or open (negatively-curved). Comparison with the BOOMERanG image (top) indicates that space is very nearly flat.

When the temperature fluctuations in the cosmic microwave background are plotted against angular size, the resulting graph (a) displays a number of peaks. (b) The first peak corresponds to clumps of matter and photons which have collapsed to maximum compression for the first time, the second to clumps that have rebounded to maximum rarefaction, and the third to clumps that have collapsed for the second time. The red curve in (a) shows the effect of including a higher proportion of baryons in the cosmic mix.

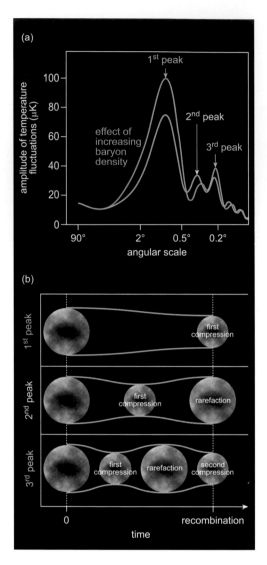

A view towards the Cosmic Background Imager site on the Llano de Chajnantor, in the Chilean Andes, photographed from Cerro Toco.

on a host of specific cosmological parameters, such as the geometry of space, the relative proportions of the different constituents of the universe, and the rate at which the universe is expanding. The total amounts and relative proportions of baryons, photons and cold dark matter particles affect the heights, shapes and spacings of the peaks in the power spectrum. For example, if the proportion of baryons and electrons is increased, the heights of the first and third peaks are amplified but the height of the second peak is reduced. Potentially, therefore, cosmologists can extract a wealth of information from the measured spectrum of temperature fluctuations.

In addition to measuring the first peak (which showed space is flat) the BOOMERanG and MAXIMA experiments also uncovered hints of the second and third peaks. Since that time, sophisticated ground-based instruments have succeeded in mapping cosmic ripples with much-improved resolution and precision, though admittedly only over fairly small areas of sky. In order to operate successfully, instruments of this kind need to be located at sites (usually at high altitudes) where the amount of atmospheric water vapour (which absorbs microwaves) is as low as possible. One of the leading ground-based experiments is the Degree Angular Scale Interferometer (DASI), an array of 13 microwave receivers which was set up by a consortium of American universities at the South Pole, where most of the atmospheric water vapour is frozen out. The Arcminute Cosmology Bolometer Array Receiver (ACBAR) is another American instrument that takes advantage of the unique south-polar environment. Other ground-based instruments include the intriguingly-named Very Small Array (VSA), a set of 14 small-scale receivers on Mount Teide on the island of Tenerife, and the Cosmic Background Imager (CBI), which is located at a height of 5080 metres (16,700 feet) on the Llano de Chajnantor, a high plateau in the Chilean Andes – so far above the bulk of the Earth's atmosphere that observers need to take oxygen cylinders with them. Experiments such as these have made, and continue to make, vital contributions to improving our detailed knowledge of the microwave background and the various peaks in its pattern of ripples.

The most extensive and comprehensive all-sky study of the microwave background so

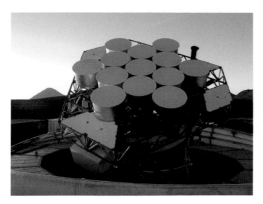

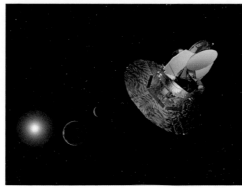

(Left) The Cosmic Background Imager, shown here, consists of 13 separate radio antennae on a single mount that can be pointed to a particular direction in space. Each antenna is a 90-cm concave reflector enclosed in a shield and protected by a Teflon cover that is transparent to radio waves.

(Right) This artist's depiction shows the WMAP spacecraft at L2 - its distant location, some 1.5 million kilometres from the Earth - with the Sun, Earth and Moon in the background.

far has been conducted by NASA's Wilkinson Microwave Anisotropy Probe (WMAP), a spacecraft that was launched on 30 June 2001 and which was named, after its launch, in honour of Princeton University cosmologist, the late David T Wilkinson, who was one of the architects of the mission. WMAP was placed at a particular stable location in space called L2 (the second 'Lagrangian point'), which is 1.5 million kilometres from Earth on the opposite side of the Earth from the Sun. At this particular location, the combined gravitational attractions of the Earth and Sun ensure that any spacecraft that is 'parked' there orbits the Sun in the same period of time as the Earth (one year) and remains at all times on the opposite side of the Earth from the Sun. WMAP carries two reflecting telescopes, mounted back to back, which look simultaneously at, and measure the temperature difference between, two widely-spaced regions of sky. The combined motion resulting from the rate at which the spacecraft spins around its axis, and its orbital motion around the Sun, enables WMAP to scan the entire sky once every six months. By comparing the measured temperature differences between all the different pairs of sky patches that the spacecraft observed, the WMAP Science Team was able to construct an all-sky map of the temperature fluctuations. WMAP is about forty-five times more sensitive than COBE was, and its angular resolution is about thirty-three times better (about 0.2 degrees compared to 7 degrees).

WMAP also has the capacity to measure polarisation. Polarisation is a property of light, or any other kind of electromagnetic wave,

which relates to the way in which light waves vibrate. Like a wave on water, a light wave vibrates perpendicular to the direction in which the wave is advancing (if a water wave is propagating in a horizontal direction, the water through which it is travelling moves up and down). Viewed from head-on, the electric or magnetic disturbance, which constitutes a light wave, vibrates in some particular direction (up-down, left-right, or whatever). If a beam of light contains waves that are vibrating equally in all possible directions, it is said to be unpolarised. If, on the other hand, all the waves are vibrating in one particular direction, the beam is said to be plane polarised. There are various possible states of polarisation that lie between these two extremes.

If an initially unpolarised beam of light (such as sunlight) reflects off a flat surface (such as a mirror, the sea, or the surface of a road) it becomes polarised in such a way that the reflected light waves are vibrating parallel to

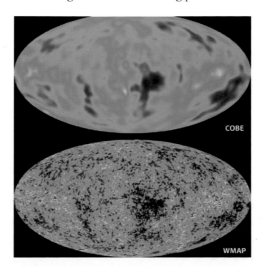

The lower of the two all-sky maps shows how the WMAP image brings the pattern of warmer and cooler spots in the microwave (the fossilised imprint of the infant universe, 380,000 years after the Big Bang) into much sharper focus compared to the COBE image (above).

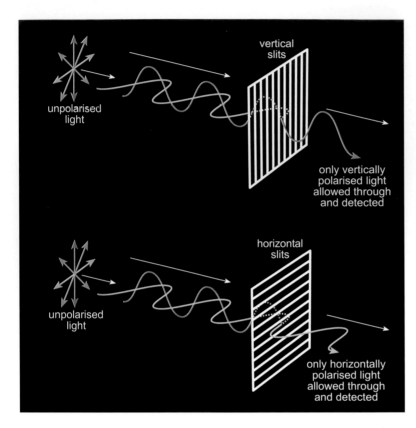

An unpolarised beam of light (left) contains waves vibrating in all different planes, whereas a plane polarised beam vibrates in one plane only (for example, vertical or horizontal). The polarisation in a beam of light can be measured by using an analyser that behaves rather like a grid of slits, and which only transmits light which is vibrating parallel to the slits.

the surface. Polarising sunglasses make use of this fact to cut out reflected glare. The material of which polarising lenses are made is rather like a grid (like a gate with vertical spars and gaps between the spars) which transmits light vibrations that are lined up with the 'gaps' in the grid, but block out vibrations that are at right angles to the grid. If you have a pair of polarising sunglasses, you can easily demonstrate the principle by looking at the reflected glare of sunlight on a surface, and then slowly rotating the glasses from horizontal to vertical. When they are horizontal (as you would normally wear them), the glare is very much reduced, whereas when they are held at right angles to the normal direction, the glare passes straight through the lenses; the polarising grid is orientated so as to cut out the waves that are vibrating in a horizontal plane (the polarised reflected light) while still transmitting direct light which is vibrating in other directions. As a result, you can still see where you are going, but you are not dazzled by the reflected glare. Because electromagnetic radiation becomes polarised to some degree when it bounces off

(is scattered by) electrically-charged particles, such as electrons, analysis of polarisation effects in the microwave background provides another rich source of information about the state of matter and radiation at the time of decoupling and about what has happened to that radiation en route to the Earth (for example, if it has passed though an intervening cloud of ionised gas).

In 2003, the Science Team published their analysis of the first year's results obtained by the WMAP spacecraft. The outcome was a veritable goldmine of science data, of which the following is only a sample. The team deduced a value for the Hubble constant of 71 km/s/Mpc (almost exactly the same as the value that had been obtained from an extensive study of Cepheid variable stars in galaxies carried out by a Hubble Space Telescope team) and established that the universe is 13.7 billion years old with a possible error of little more than 1 percent. They showed that space became transparent (matter and radiation decoupled) 380,000 years after the beginning of time and that the interval of time over which the process of decoupling was spread was about 118,000 years. By measuring the overall degree of polarisation in the background radiation, and assuming that this was caused by clouds of electrons that were produced when the first stars and luminous objects began to shine, and thereby began to reionise some of the hydrogen atoms that had formed at the time of recombination, the team deduced that the first highly-luminous stars had formed and begun to shine just 200 million years after the Big Bang. This figure was considerably earlier than most theoreticians had expected, although in 2006 it was revised to 400 million years (see Chapter 12). This result itself effectively ruled out 'hot', or 'warm' dark matter as a major constituent of the universe[2]; hot or warm dark matter would have been moving so fast that gravity would not have been able to pull matter together to create objects such as these so early in the history of the universe.

WMAP confirmed (as indeed had earlier investigations by COBE, BOOMERanG, MAXIMA and various ground-based experiments) that the detailed properties of the observed warmer and cooler patches, measured over a wide range of angular sizes, matched rather well with the pattern of fluctuations that had been predicted by inflationary theory. This provided further support for the flat-space, critical density, inflationary model.

The WMAP team's detailed analysis of the power spectrum and its various peaks confirmed that the first and highest peak occurs at an angular scale of about one degree, and that space, therefore, is flat. Their calculated value for the mean overall density was almost exactly equal to the critical density. Analysis of the heights, locations and shapes of the peaks indicated that the total matter density in the universe is about 27 percent of the critical density, and the ratio of the heights of the first and second peaks showed that the total amount of baryonic matter is 4.4 percent of the critical density – in very close agreement with what is predicted from studies of Big Bang nucleosynthesis (see Chapter 5). The overall density of cold dark matter, therefore, is equivalent to about 23 percent of the critical density. The detailed appearance of the peaks and troughs also places limits on any contribution to the overall density by neutrinos (hot dark matter); the WMAP results show the total amount of mass contained in the neutrino population adds up to less than 1.5 percent of the critical density (and probably considerably less).

A useful shorthand is to express the contribution of each of these various constituents as a fraction of the critical density – an 'Omega' value for each. Expressed in these terms, the WMAP team's 2003 results are as follows: $\Omega_{total} = 1.02 \pm 0.02$, Ω_M (the total matter fraction) $= 0.270 \pm 0.016$, Ω_b (the baryon fraction) $= 0.044 \pm 0.004$, Ω_v (the neutrino fraction) < 0.015; the deduced cold dark matter fraction is $\Omega_{CDM} = 0.23$.

Three years later, in March 2006, the WMAP Science Team published the results that had been obtained during the first three years of the spacecraft's mission. Their analysis of the data produced more precise values of the data (see Chapter 12) which nevertheless differed only slightly from the 2003 results. In round fig-

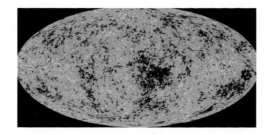

ures, the three-year results gave the following figures for the relative proportions of the major forms of matter: total matter, 26 percent ($\Omega_M = 0.26$); baryonic matter, 4 percent ($\Omega_b = 0.04$); cold dark matter 22 percent ($\Omega_{CDM} = 0.22$).

Corroboration from matter clustering

The CMB provides a snapshot of the way in which matter was distributed 380,000 years after the beginning of time. But astronomers have the means to acquire snapshots at more recent times, too. Large-scale galaxy redshift surveys, such as 2dFGRS and SDSS (see Chapter 3), have already succeeded in mapping the distribution of hundreds of thousands of galaxies out to distances of several billion light-years, which corresponds to looking at the large-scale distribution of luminous matter in recent times (in our local part of the universe) and as far back as several billion years ago. Furthermore, by analysing the large numbers of absorption lines imprinted on the spectra of remote quasars (the so-called 'Lyman-alpha forest') by numerous intervening clouds of hydrogen gas, astronomers can map the distribution of clouds of hydrogen atoms at redshifts in the region of 2–4, which corresponds to viewing the distribution of matter 1–3 billion years after the Big Bang. In addition, large-scale surveys of so-called 'weak' gravitational lensing (see Chapter 12) hold out

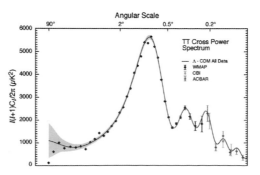

This first detailed all-sky picture of the infant universe, produced by the WMAP spacecraft, reveals temperature fluctuations (warmer and cooler patches shown as colour differences) that correspond to the seeds from which galaxies and clusters grew. Encoded in the patterns is a wealth of information about fundamental properties of the universe.

The 'angular power spectrum' of the temperature fluctuations in the WMAP full-sky map plots the relative brightness of the 'spots' on the map against the angular size of the spots. Data from other experiments are also included. The shape of the curve is a 'fingerprint' which contains a wealth of information about the age, content and history of the universe.

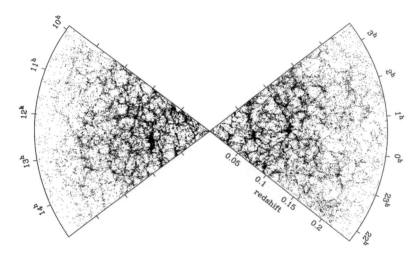

This sample of data from the Two-Degree Field Galaxy Redshift survey (previously encountered in Chapter Five) shows the large-scale distribution of luminous matter in the universe. The structures revealed here are believed to have evolved from localised primordial density differences associated with the temperature fluctuations that have been detected in the cosmic microwave background.

the promise of being able to map the total distribution of matter (both luminous and dark) over a similar range of redshifts.

The distribution of gas clouds in the more distant past and the way in which galaxies and clusters are organised in more recent times reflects the way in which the distribution of matter has evolved over time from the underlying density fluctuations that were imprinted on the CMB. By comparing the patterns of structure, of various sizes, that surveys such as these reveal in the distribution of galaxies and matter with those which are predicted by theoretical models with different amounts and relative proportions of baryons and dark matter, cosmologists can deduce values for a range of cosmological parameters such as: the overall matter density; the baryon density; and the curvature (or flatness) of space. Despite some

detailed differences, the bottom line is that all of these separate approaches point towards the same general 'concordance' scenario: an inflationary universe with essentially flat-space, with an overall density of matter and energy equal to, or exceedingly close to, the critical density, and within which the overall matter density is about 26 percent of critical, the baryon density is 4–5 percent of critical, and the density of cold dark matter is about 22 percent of critical[3].

If space is flat and the density is critical ($\Omega_{total} = 1$), but matter in all its various guises adds up to just 26 percent of the critical density ($\Omega_m = 0.26$), then simple arithmetic shows that 74 percent of the overall density of the universe must be contributed by 'something else', the nature of which is as yet unknown. From Einstein's theory of relativity, we know that matter and energy are equivalent – a quantity of energy is equivalent to a quantity of mass, so that energy, too, has a gravitational influence. The missing link in the cosmic balance is dark, in the sense of not being 'visible' in any direct way. It is *not* matter, in the form of baryons or non-baryonic particles. It may instead be a hitherto unknown form of energy – an invisible distribution of energy that permeates the universe but does not clump together in the way that matter does. The missing commodity has come to be known as 'dark energy'.

But, even before BOOMERanG, MAXIMA and WMAP had compelled astronomers to grasp the nettle and treat seriously the idea that dark energy is the dominant component of the cosmos, the need for its existence had already been demonstrated by a completely different, and (for many astronomers) wholly unexpected, discovery involving supernovae (exploding stars), the story of which will unfold in Chapter 9.

This 'pie' diagram displays the content of the universe as revealed by the three-year WMAP results. The various slices indicate how much of the overall density of the universe is contributed by ordinary matter ('atoms'), cold dark matter and dark energy.

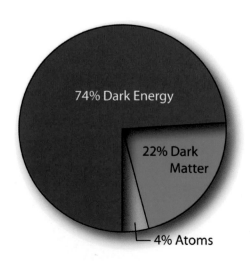

74% Dark Energy

22% Dark Matter

4% Atoms

Runaway Universe:

Exploding Stars Point to Accelerating Expansion

Until the late 1990s, most cosmologists assumed that the gravitational influence of matter is the dominant force in the universe and that the rate of cosmic expansion has been slowing down continuously since the Big Bang, or, at least, since the brief interlude of accelerating expansion that took place during the inflationary era. But if, as the growing body of observational evidence seemed to indicate, the mean matter density was in the region of 20–30 percent of the critical density, then the universe would be open, gently decelerating, and fated to expand forever.

As an alternative to trying to measure the mean density directly, some cosmologists focused their attention instead on trying to measure changes in the expansion rate over the history of the universe. In principle, astronomers can do this by measuring the distances and redshifts of a suitable type of 'standard candle' (an object of known intrinsic luminosity which is sufficiently brilliant to be visible at enormous distances) over as wide a range of distance and redshift as possible. As we saw in Chapter 2, the redshift (z) of a remote object depends on the amount by which the universe has expanded between the instant at which light departed from that object and the instant at which it reached the Earth (i.e. the present time). To be precise, the ratio of the scale of the universe at the time light was emitted (R_e) to its scale at the present time (R_0) is equal to $1/(1 + z)$. For example, if the redshift in the spectrum of a distant source is 1 ($z = 1$), the expansion of space has stretched that light to twice its original wavelength, and the scale of the universe (the separation between galaxies) today is twice as great as it was at the instant when the light we are now receiving departed from that object. We are seeing that object as it was when the scale of the universe was half what it is now ($1/(1 + z) = 1/(1 + 1) = \frac{1}{2}$). Redshift, then, tells us the scale of the universe at the time when light departed from our chosen standard candle.

The more distant an object is, the fainter it will look. By comparing a standard candle's observed apparent brightness with its known (or assumed) luminosity, we can work out how far away it must be in order to appear as faint as it does. This measure of distance is called luminosity distance (see Chapter 1). Knowing an object's distance, we can work out how long its light has taken to reach us. Distance tells us how far back in time we are looking, and redshift tells us the scale of the universe at that time. If we measure the apparent brightness and redshift of a large enough number of standard candles, spread over a wide enough range of distances, we can find out how (or if) the rate of expansion has changed with time, and can piece together the entire expansion history of the universe.

The quest for an ideal standard candle

In order to achieve this goal, astronomers need to find the right kind of standard candle. The ideal standard candle would be a set of objects which are extremely luminous (so as to be detectable at very great distances) and which have precisely the same light output (like identical light bulbs with the same wattage).

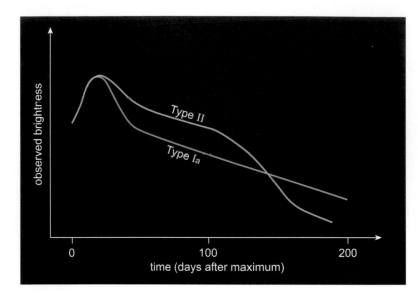

Typical light curves (graphs of brightness versus time) for a Type II supernova and a Type Ia supernova.

out that supernovae – exploding stars which for a time may become as bright as an entire galaxy – could well be extremely useful for distance measurement. Unfortunately, as more supernovae were discovered and investigated, it appeared that they had a wide and diverse range of properties and a broad spread of peak luminosities – not at all what was needed for a standard candle.

Supernovae are divided into two broad classes – Type I and Type II – based on their spectra and light curves (graphs of how their brightness rises to a maximum then fades away). Whereas the spectrum of a Type II supernova exhibits prominent hydrogen lines, the spectrum of a Type I supernova does not. In the 1980s, Type I supernovae were further subdivided into Type Ia and Type Ib (and subsequently, Type Ic), depending on whether or not their spectra included a strong absorption line at a wavelength of 615 nanometres due to the element silicon (Type Ia supernovae show this feature, but Types Ib and Ic do not). It

The perfect standard candle does not appear to exist in nature, which is why the measurement of large cosmic distances has always been one of the most difficult tasks in observational astronomy. However, as early as 1938 Walter Baade and Fritz Zwicky (again!) had pointed

The Crab nebula, shown here in a mosaic image taken by the Hubble Space Telescope, is the six light-year wide remnant of a supernova which was seen by Chinese and Japanese observers in the year 1054 and which marked the explosive demise of a high-mass star.

then became apparent to astronomers that all Type Ia supernovae were closely similar – they all had very similar spectral features, their light curves were closely similar, and they all attained very similar peak luminosities – about 4 billion times the luminosity of the Sun.

The differences between the various supernova types reflect the nature of the process that detonates these cosmic conflagrations. A Type II supernova is a high-mass star (many times more massive than the Sun) which burns a succession of nuclear fuels until its core is converted into iron. The core then collapses to form an exceedingly dense and compact neutron star (see Chapter 1), and the remainder of the star's material is blasted violently into space. The expanding remnant includes the star's outer envelope of unconsumed hydrogen, which gives rise to the observed hydrogen lines in the supernova's spectrum. Supernovae of Types Ib and Ic are now believed also to be 'core-collapse' supernovae, but involving high-mass stars that lose their outer envelopes of hydrogen before the detonation occurs and which, therefore, do not have hydrogen lines in their spectra.

Astronomers believe that Type Ia supernovae are the result of a completely different process. An event of this kind is believed to involve a close binary system (a binary consists of two stars that revolve round each other) which consists of a white dwarf – the shrunken remnant of an old Sun-like star – and a large distended star (such as a red giant) that is still generating energy by means of nuclear fusion reactions. If the two stars are sufficiently close together, gas will stream from the larger companion down on to the surface of the white dwarf, accumulating there and gradually increasing that compact star's mass. A white dwarf can only support itself against the inward pull of gravity if its mass is less than a certain critical value (around 1.4 solar masses), which is called the Chandrasekhar limit. If the steady accretion of gas pushes the white dwarf's mass up to this limit, its internal pressure will be overwhelmed by gravity, and it will begin to collapse. However, white dwarfs contain large amounts of potential nuclear fuel, such

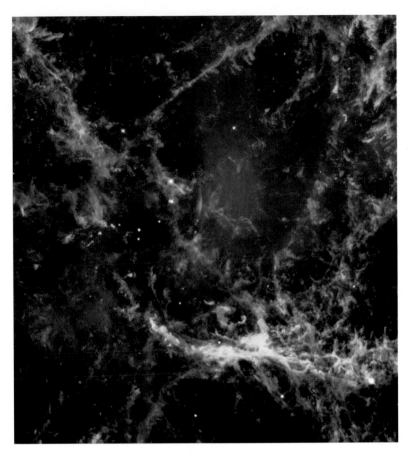

as carbon and oxygen which, hitherto, have not taken part in energy-releasing fusion reactions because the star's internal temperature has not been high enough. As soon as the star begins to collapse, the temperature rockets up, triggering a sudden and catastrophic burst of carbon and oxygen 'burning' – a violent release of thermonuclear energy that blows the star apart.

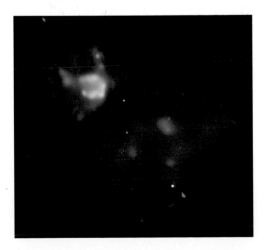

In this close-up image of the heart of the Crab nebula, the lower of the two moderately bright stars to the upper left of centre is the compressed core of the exploded star, which has survived as an exceedingly compact, rapidly rotating neutron star.

This composite x-ray (red and green) and optical (blue) image shows apparently adjacent remnants of two supernovae in the Large Magellanic Cloud. The hot shell on the upper left is the remnant of a Type Ia supernova whereas the one on the lower right is a Type II remnant. Their apparent proximity is probably due to a chance alignment.

This sequence of stages shows the mechanism of a Type Ia supernova. If the accumulation of gas, which has been flowing from a companion star, pushes the mass of a white dwarf over the maximum permitted limit, a catastrophic explosion ensues.

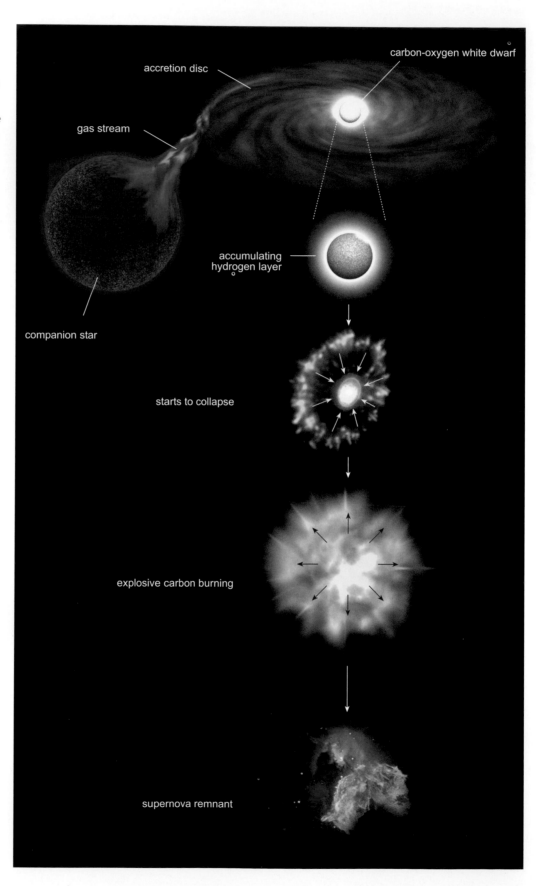

carbon-oxygen white dwarf

accretion disc

gas stream

accumulating hydrogen layer

companion star

starts to collapse

explosive carbon burning

supernova remnant

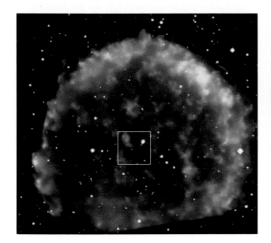

These images show the location of the suspected companion star to the supernova that was seen in the year 1572 by the Danish astronomer, Tycho Brahe. The image on the right, which shows the candidate star (labelled 'Tycho G'), has been superimposed (left) on an x-ray image of the supernova remnant. The location and speed of the star is consistent with its having been catapulted away when its erstwhile companion exploded. This discovery provides the first direct evidence to show that Type Ia supernovae come from binary star systems containing a normal star and a white dwarf.

Because these cataclysmic events involve the total destruction of stars with near-identical masses, all Type Ia supernovae are expected to release closely similar amounts of energy and to rise to closely similar peak brilliancies. A typical Type Ia supernova reaches a peak luminosity about 4 billion times that of the Sun, which is brighter than the entire light output of many galaxies. Because of their great brilliance and the close similarity between their peak brilliancies, these high-luminosity cosmic beacons come closer to approaching 'ideal standard candle' status than anything else within the observational cosmologist's toolkit.

In practice, there *are* some differences between individual Type Ia supernovae, but by accumulating data on large numbers of relatively nearby supernovae, astronomers have been able to build up standard 'templates' of their light curves and spectra against which newly-discovered supernovae can be calibrated – to determine their peak brilliancy and to weed out anomalous ones. Astronomers have found that the detailed shape of a Type Ia supernova's light curve is related to its peak brightness and that, in particular, supernovae that brighten and fade more rapidly than the norm are less luminous than average. In addition, by measuring the light curve through different colour filters, astronomers can detect the extent to which the supernova's light has been dimmed and 'reddened' by passing through clouds of dust in our Galaxy or its host galaxy. Because particles of dust scatter and absorb light, a foreground cloud of dust causes

a background star to appear fainter than it otherwise would do; furthermore, because dust scatters short-wave blue light more than long-wave red, the blue component of a star's light is depleted more than the red component, with the result that the star looks 'redder' than it otherwise would. Taking these factors into account, astronomers are reasonably confident about deducing the intrinsic peak luminosity of each Type Ia supernova that they find in a distant galaxy.

Whereas Type Ia supernovae clearly have great potential as distance indicators, they have drawbacks too. First, Type Ia supernovae are very rare events: on average, within a typical galaxy, only two or three will occur in 1000 years. Secondly, they are unpredictable: astronomers have no advance warning of when or where the next supernova is going to occur. Thirdly, they are short-lived phenomena: they have to be discovered promptly, and measured repeatedly over a matter of just a few weeks, otherwise astronomers may not be able to

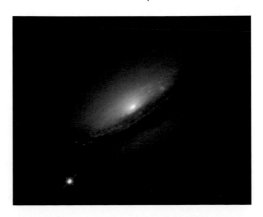

Supernova 1994D, the bright point of light towards the lower left, was a Type Ia supernova that occurred in galaxy NGC 4526, which is a member of the Virgo clusters of galaxies, and is located at a distance of about 50 million light-years. At maximum brightness it was as luminous as 4 billion suns.

see them at peak brilliancy, or may miss them altogether. Whereas Type Ia supernovae can be discovered with medium-sized (or even quite modest) telescopes, astronomers need very large instruments to capture enough light to produce spectra and to carry out precision photometry (measurements of brightness) as the brightness rises and falls. But there are many demands on, and great competition for, observing time on the largest telescopes. As a result, slots of observing time are allocated to different teams well in advance as part of a carefully compiled schedule. The largest telescopes cannot normally be diverted at the drop of a hat to concentrate on a supernova that has suddenly appeared, nor can they be tied up looking at some remote galaxy on the off-chance that a supernova might suddenly flare up – that would not be a cost-effective use of precious telescope time.

Despite these difficulties, a number of observational cosmologists became convinced that Type Ia supernovae had such great potential for probing the expansion history of the universe that they set about evolving strategies to make it all happen. By the mid-1990s, two international research teams – the Supernova Cosmology Project (SCP), headed by Saul Perlmutter of the Lawrence Berkeley Laboratory, California, and the High-Z (High-redshift) Supernova Search Team (HZT), headed by Brian Schmidt of the Mount Stromlo and Siding Springs Observatories in Australia – had begun systematic searches for remote Type Ia supernovae in an attempt to pin down various cosmological parameters, including the deceleration rate, the Hubble constant, the mean matter density and the overall mass-energy density of the universe.[1]

Both teams developed a strategy – first demonstrated in 1994 by Perlmutter's team – that in effect *guarantees* finding batches of distant supernovae during allocated sets of observing nights, an approach that allows follow-up observations with very large telescopes to be scheduled in advance. The first step is to use wide-field cameras on large (typically around 4-metre aperture) telescopes to take a set of images of adjacent patches of sky containing tens of thousands of galaxies around, or just after, new Moon, when the sky is as dark as possible. A second set of images of these same regions of sky is taken three to four weeks later, just before the next new Moon. The two sets of images are then immediately compared, to pick out any new bright points of light that have appeared, some of which will be Type Ia supernovae that have flared up during the time interval between the two sets of exposures. The selected patches of sky contain so many galaxies that the search teams can be virtually certain of finding a dozen or so new supernovae in the second set of images. During the dark nights around the second new Moon, pre-scheduled follow-up observations are made, using very large instruments such as the 10-metre Keck telescopes and 8-metre Gemini, Subaru and VLT (Very Large Telescope) instruments on high mountain sites in Hawaii and Chile, or the Hubble Space Telescope in Earth orbit, to obtain spectra which are used to confirm which of the objects are indeed Type Ia supernovae, to determine their redshifts, and to plot their light curves.

The measured peak brightnesses of Type Ia supernovae are then plotted against redshift to yield a graph, called a Hubble diagram, which can be compared with theoretically predicted curves for model universes with different values of the key cosmological parameters – in particular, the Hubble constant, mean matter density, total mass-energy density, and spatial curvature. Because each of these quantities affects the observed brightness and implied distance of a supernova (or anything else), the exact shape of the Hubble diagram depends on the precise values of each of these parameters.

In the mid-1990s, the prevailing opinion was that the evolution of the universe is controlled by the gravitational influence of matter alone and that the major constituent of the present-day universe is cold dark matter. If the popular inflationary model were valid, the mean density of the universe would be equal to the critical density ($\Omega = 1$) and the net curvature of space would be zero. The general expectation was that the rate of expansion should be slowing down in the same sort of way as would be anticipated for a flat 'Einstein–de Sitter' universe – a universe that could just, but only just, continue to expand forever. The supernova teams were expecting to determine the rate at which the expansion was slowing down (the 'deceleration parameter').

A decelerating universe would have been expanding faster in the past than it is now. Consequently, galaxies would have taken less time to reach their present distances than would have been the case had the universe been expanding at a constant rate (a so-called 'coasting' universe). For example, suppose that a supernova is embedded in a galaxy with a redshift of 0.5. Light from that supernova was emitted when the universe was two-thirds of its present size and would have been travelling through space (at a constant speed) for a shorter time in a decelerating universe than in a coasting universe. That supernova would be nearer to us – and hence brighter – than the equivalent supernova in a coasting universe. Conversely, if the expansion were accelerating, the universe would have been expanding more slowly in the past than it is now. Galaxies would have taken longer to reach their present distances than would have been the case in a coasting universe and light from a supernova at a particular redshift would have travelled further and for longer in order to reach us. That supernova would be further away – and hence fainter – than the equivalent supernova in a coasting universe. The amount by which the observed Hubble diagram deviates from the Hubble diagram for a coasting universe depends on the rate at which the expansion is slowing down (or speeding up).

Results for the first few supernovae began to emerge in 1997, and were rather indeterminate. At the beginning of that year, the Supernova Cosmology Project published results, based on their first seven supernovae, which implied a high value for matter density, broadly consistent with a universe decelerating in similar fashion to an Einstein–de Sitter model, but much larger than estimates of the matter density that had been obtained by other means. But by adding just one more supernova later in the year (supernova SN 1997ap, which,

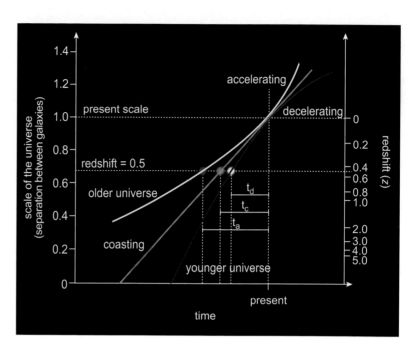

with a redshift of 0.83, was the most distant that had been discovered at that time) the SCP team found that they had to revise their figure for the matter density quite substantially downwards. In the meantime, the High-Z team, using four supernovae out to a maximum redshift of 0.97 (a new redshift record), also came up with relatively low figures for the overall matter density (which implied that at most, $\Omega_M = 0.35 \pm 0.3$). With all of these early results, the uncertainties in the measurements were too great to allow any very firm conclusions to be drawn, but on balance the first sets of data seemed to be pointing towards a gently decelerating low matter density universe.

Supernovae deliver a shock result

But in 1998, with the benefit of improved analysis and more supernovae, both teams came up with a truly startling result. They each found, independently, that Type Ia supernovae in high-redshift galaxies were systematically fainter, and hence further away, than would be expected if the universe were coasting or expanding at a decelerating rate. Specifically, they found that high-redshift supernovae (at redshifts of around 0.5 and look-back times of around 6 billion years) were about 15 percent fainter than would be expected in a coasting universe, and about 20–25 percent fainter than

When we look at a supernova embedded in a galaxy at redshift z = 0.5 we are looking back to when the universe was two thirds of its present scale. This diagram shows how the time for which light has been travelling from, and hence the distance and observed faintness of, a supernova depends on whether the universe is expanding at a decelerating, constant or accelerating rate.

would be expected in a decelerating universe in which matter with a density equal to about 30 percent of the critical density ($\Omega_M = 0.3$) was the sole ingredient. These dramatic and unexpected results suggested that, far from slowing down under the influence of gravity, the expansion of the universe is actually *accelerating*.

At the January 1998 American Astronomical Society meeting in Washington DC, the SCP team, which by then had accumulated data on 42 supernovae, gave a tentative hint that their data might possibly be pointing towards accelerating expansion. Then, at a dark matter conference held in Los Angeles in February of that year, the High-Z team reported results for 16 high-redshift supernovae which, they contended showed that the universe is accelerating and that a significant component of its overall density is made up of some kind of (what is now known as) dark energy. The High-Z team's results were published in *Astronomical Journal* in September 1998. Intriguingly, their analysis yielded a cosmic matter density (Ω_M) of 0.24 ± 0.10 if the overall mass-energy density were equal to the critical density (i.e. $\Omega_{total} = 1$), but if the dark energy density were taken to be zero, as was widely assumed prior

to that time, then the value of the matter density came out as $\Omega_M = -0.35 \pm 0.18$ – a *negative* and completely unphysical value. Without including acceleration (and dark energy) the results appeared to make no sense at all.

The Supernova Cosmology Project team's full analysis of their set of 42 high-redshift supernovae was submitted to *Astrophysical Journal* in September 1998 and published in June 1999. Their results pointed to the same conclusion as had been arrived at by the High-Z team, namely, that the universe is expanding at an accelerating rate and that some form of dark energy currently outweighs matter in all its forms (luminous or dark). The close agreement between the results obtained by two independent groups, based on largely independent sets of supernovae (there were only two supernovae in common between the two sets), was truly remarkable and compelled the scientific community to treat the evidence for accelerating cosmic expansion very seriously indeed. Indeed, in its issue of 18 December 1998, the prestigious American journal *Science* hailed the discovery by the HZT and SCP of the accelerating universe as the top 'Science Breakthrough of the Year'.

The evidence for accelerating expansion has become more and more persuasive as increasing numbers of remote supernovae have been discovered and the quality of the measurements has improved. By 2003, when John L Tonry, of the University of Hawaii, and other members of the High-Z Supernova Search Team drew together all the data sets that were available at that time – including observations made with the Hubble Space Telescope and a host of ground-based instruments – they were able to compile a tally of distance measurements for no less than 230 Type Ia supernovae with redshifts of up to around $z = 1$. From this database they were able not only to confirm with confidence that the expansion of the universe is indeed accelerating, but also to calculate figures for the matter density and dark energy density comparable in precision and value to those obtained, from studies of the cosmic microwave background, by the WMAP spacecraft (see Chapter 8).

The observed apparent magnitude (a measure of relative brightness) versus redshift is plotted here for nearby and distant Type Ia supernovae. When compared with theoretical predictions (shown by the curves) for different densities of matter and vacuum energy (dark energy), the best fit to the measured data points (the blue curve) matches an accelerating universe in which matter contributes about one third of the density and dark energy, two thirds.

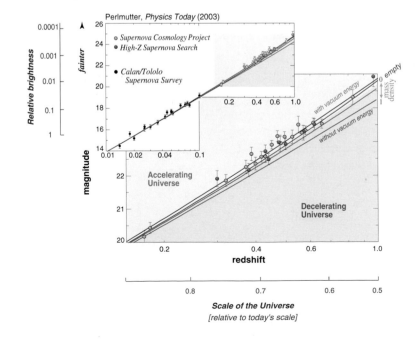

Type Ia Supernovae

Later in that same year, R A Knop and the Supernova Cosmology Project team published high quality data on 11 supernovae that they had studied with the Hubble Space Telescope. The quality of their brightness measurements and light curves, coupled with improved colour measurements that allowed the team to calculate the extent to which supernovae had been dimmed by the obscuring effects of clouds of dust (dust extinction) within their host galaxies, was such that this set of results on its own was good enough to confirm – independently of all previous results – the acceleration of the universe and the need for dark energy with a statistical confidence of better than 99 percent. Since that time, the number and quality of supernova detections has continued to grow, and the case for accelerating expansion has become progressively more compelling.

Real or apparent?

When the claims of accelerating expansion were unleashed on the astronomical community, many cosmologists were at first cautious and sceptical (as indeed had been the members of the SCP and High-Z teams themselves). Were there other factors which could cause remote supernovae to appear dimmer than expected – effects which could mimic those which had been interpreted as acceleration? The discovery of accelerating expansion would have such profound implications for physics and cosmology that astronomers need to be certain that they have eliminated all other possibilities.

One possibility is that the underlying assumption that all Type Ia supernovae have the same peak luminosity may be unsound. Perhaps more distant supernovae – which are viewed as they were in the distant past – appear less luminous because supernovae *were* less luminous in the past than they are now. After all, because nuclear reactions which take place within stars synthesise quantities of heavy elements (elements heavier than hydrogen and helium), a proportion of which is ejected into space towards the ends of their lives, the chemical composition of the clouds

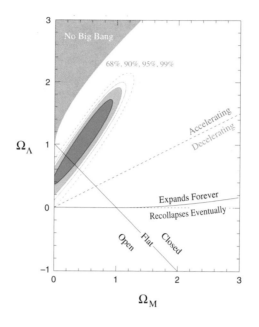

The various ellipses show the constraints (with different levels of confidence) placed on matter density (Ω_M) and dark energy density (Ω_Λ) by one set of supernova observations obtained by members of the Supernova Cosmology Project. Combining these results with the assumption of flat space gives an Ω_M value of about 0.25 and an Ω_Λ value of about 0.75.

of gas within which new stars (including those that eventually become Type Ia supernovae) are born gradually changes over time. The resulting changes in the proportion of heavier elements (what astronomers call *metallicity*) contained within successive generations of stars may have produced systematic changes in the properties of supernovae over cosmic timescales, with the result that supernovae which exploded in recent times differ subtly from those that exploded in the distant past. Or again, if the initial masses (the 'birth weight') of successive generations of progenitor stars (the stars which eventually become Type Ia supernovae) gradually change as galaxies evolve, the resulting white dwarfs might end up with systematically different proportions of carbon and oxygen – which could modify the amount of energy released in Type Ia events.

Astronomers have tried to check out possible evolutionary effects in various ways. They have compared the characteristics of large numbers of Type Ia supernovae in a wide range of host galaxy types to see if factors such as age, metallicity and the general environment within which supernovae occur have any systematic influence; they have found marginal differences, but no evidence of any really significant effects. Astronomers have also compared high-redshift supernovae to low-redshift ones and have found that their spectra are closely similar – which suggests that Type Ia supernovae were pretty much the same in the past as they are now. They have looked at whether or not there are any significant differences in the light curves of distant and nearby supernovae. While there is some evidence to suggest that high-redshift supernovae rise to peak brightness a little faster than low-redshift ones, there is no clear evidence to suggest that this affects their peak luminosities. On balance, there are no clear indications that the evolution of Type Ia supernovae over cosmic time in any significant way undermines the luminosity distance data.

Another effect to be considered is gravitational lensing. While travelling vast distances through space, light from supernovae will be focused and defocused by the gravitational influence of large-scale distributions of mass (for example, galaxy clusters), and by micro-lensing caused by compact objects such as stars or MACHOs. The cumulative effect is to *reduce* the apparent brightness of very remote objects, but estimates suggest that, at most, lensing effects should make only a rather small contribution to the observed 25 percent dimming of supernovae.

Dust is another potential problem. We know that our own Galaxy contains a tenuous distribution of tiny solid particles, called interstellar dust, which absorbs and scatters starlight and causes distant stars to appear fainter than they really are. However, as we have seen, these microscopic particles affect short wavelength light (blue) more than long wavelength light (red), causing distant stars to appear redder (or 'less blue') than their true colours. Because the diminution of brightness due to these tiny dust grains depends on wavelength, astronomers can detect, and compensate for, the effects of dust in our own Galaxy and in the host galaxies of supernovae by measuring their brightness through different colour filters.

However, so-called 'grey dust' – larger-sized particles that would affect all wavelengths of light equally (or nearly so) – would make distant objects appear fainter without leaving an obvious telltale imprint on their light. There is no evidence for the existence of grey dust in host galaxies (it would exert subtle measurable effects which have not been detected), but if intergalactic space (the space between the galaxies) contains grey dust, that could pose a much more serious problem. Intergalactic grey dust would cause more distant supernovae to appear progressively fainter – the more distant the supernova, the greater the absorption of light. If the dust were truly 'grey', its effects would be extremely difficult to detect. But if it were to exist in sufficient amounts to make high-redshift supernovae appear as faint as they are seen to be, that would require rather large amounts of the universe's budget of heavy elements to be tied up in this form. On balance, there is no tangible evidence to suggest that intergalactic grey dust exists, or if it does, for there to be enough of it to pose a real problem for the accelerating expansion interpretation of the faintness of remote supernovae.

A smoking gun?

Arguably the most decisive way to distinguish between genuine acceleration and cumulative systematic effects, such as absorption of light by grey dust or evolutionary changes in the properties of supernovae, is to measure how the peak brightness of supernovae deviates, at progressively higher redshifts, from what would be expected in a gently decelerating low mass-density universe with no dark energy. If – as all the evidence implies – the universe began in a hot Big Bang then, for much of its early history the mean density of matter would have been much greater than it is now; gravity would have had the upper hand, and the universe would have been decelerating. The transition from deceleration to acceleration would have occurred at some relatively recent time, when the effects of dark energy began to win out over the weakening influence of gravity. By seeking out supernovae at higher redshifts, astronomers ought to be able to look far enough back in time to view the era of decelerating expansion.

Around the transition redshift (which cosmologists anticipated would probably lie somewhere between $z = 0.5$ and $z = 1$), the deviation between the observed brightness of supernovae and the brightness that they would be expected to have in a coasting universe, or a low density universe with zero dark energy, would reach a maximum; beyond that particular redshift, the deviation would begin to decrease. Supernovae at redshifts substantially higher than this 'turnover' point (those which had exploded well back into the deceleration era) would begin to appear *brighter* than would be the case in an accelerating or coasting model. By contrast, if the deviation were due to the cumulative effects of grey dust, or the evolution of supernova properties, it should continue to increase at progressively higher redshifts, distances and look-back times. The detection of a turnover – a decrease in the brightness deviation at high redshifts – would be a clear signature of the switch from early deceleration to later acceleration and a vindication of the interpretation of supernova faintness as being due to the effects of acceleration rather than anything else.

The first strong hint that the anticipated turnover was indeed present came in April 2001, when a team headed by Adam Riess, of the Space Telescope Science Institute (STScI) reported on the discovery of what was most probably a Type Ia supernova at the startlingly-high redshift of 1.7 – the most distant supernova ever seen. The discovery owed much to an extraordinary sequence of good fortune. In December 1997, Ronald L Gilliland, of the STScI, and Mark Phillips, of the Carnegie Institutions of Washington, used the Wide-Field Camera (WFPC2) on the Hubble Space Telescope to re-image a region of sky called the Hubble Deep Field North, which had been imaged to great depth (extreme faintness) by the orbiting telescope two years previously. The aim was to try to find remote, high-red-shift supernovae by comparing the two sets of images, to see if any of the faint galaxies had changed in brightness between the two observing times. The technique involved subtracting one image from the other so that if no change had occurred the galaxy image would disappear, whereas if a change *had* occurred (if

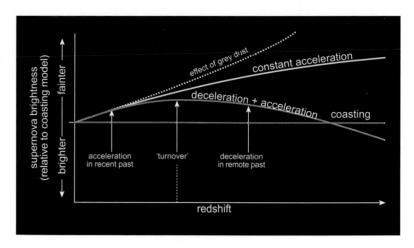

a supernova had exploded in that galaxy), the difference in brightness would show up as a blob of light. They found two good supernova candidates, one of which – the more distant of the two – was designated SN 1997ff.

The host galaxy for SN 1997ff was a faint elliptical. The galaxy's redshift had already been estimated by comparing its brightness as measured through four different colour filters in the optical part of the spectrum, but in order to get a reasonably reliable redshift from photometry (measuring brightness at a limited number of wavelengths) astronomers needed also to have measurements in the infrared part of the spectrum. As luck would have it, another team had also investigated the section of the Hubble Deep Field that contained SN 1997ff at infrared wavelengths, using another HST instrument called NICMOS (Near Infrared Camera Multi-Object Spectrograph) as part of a different research programme and, by sheer good fortune, one of their preliminary test exposures had been taken within a few hours of the image that had first revealed the supernova. Embedded within the series of NICMOS observations, and another set that was acquired some six months later, was a wealth of data relating to the supernova and its host galaxy.

When Riess and his team drew together and analysed all the available data, they were able to produce a light curve for the supernova and work out its redshift. While there was no absolute proof that SN 1997ff was definitely a Type Ia supernova (as against a Type II), the

The curves show schematically how the apparent brightness of supernovae at different redshifts would differ from the brightness of supernovae in a 'coasting universe' (one that expands at a steady rate). The three scenarios are: if the universe were undergoing constant acceleration; if it had switched from early deceleration to more recent acceleration; or if the absorbing effect of grey dust were predominant.

Supernova SN 1997ff is the most distant supernova yet seen. The small box in the top frame identifies the location of the host galaxy in the Hubble Deep Field. The galaxy is picked out by the arrow in the enlarged box (bottom left). The picture at the bottom right shows the difference between images of this region taken in 1995 and 1997; the supernova, which flared up in 1997, appears as a white dot.

light curve and colour data matched well with that possibility. The redshift turned out to be 1.7 – a record then, and still so. Crucially, its observed peak brightness turned out to be markedly *brighter* (by about 50 percent, though with a considerable margin of error) than would be the case in a coasting universe, and more than 250 percent brighter than would be expected if grey dust, evolution, or some other such factor were affecting the observed brightness of supernovae.

While this result was striking and spectacular, cosmologists remained cautious, not wishing to draw firm conclusions from a single supernova. What were needed were more discoveries and measurements of supernovae with redshifts in excess of $z = 1$, and the supernova search teams duly embarked on programmes to find them. In the spring of 2003 a team headed by John P Blakeslee, of Johns Hopkins

University, Baltimore, announced the discovery of two supernovae which they had uncovered with the (then) newly-installed Advanced Camera for Surveys (ACS) on the Hubble Space Telescope. The spectroscopic data confirmed that both were indeed of Type Ia, and their redshifts turned out to be 0.47 (in the case of SN 2002dc) and 0.96 (SN 2002dd). When their peak brightnesses were plotted against redshift, both points lay close to the predicted curve for a universe with 30 percent matter and 70 percent dark energy, which had switched from deceleration to acceleration at a redshift intermediate between the measured redshifts of these two supernovae.

In order to clarify the early expansion history of the universe, Adam Riess and co-workers began a space-based quest called the 'Hubble Higher-Z Supernova Search' to discover and monitor more supernovae at redshifts in

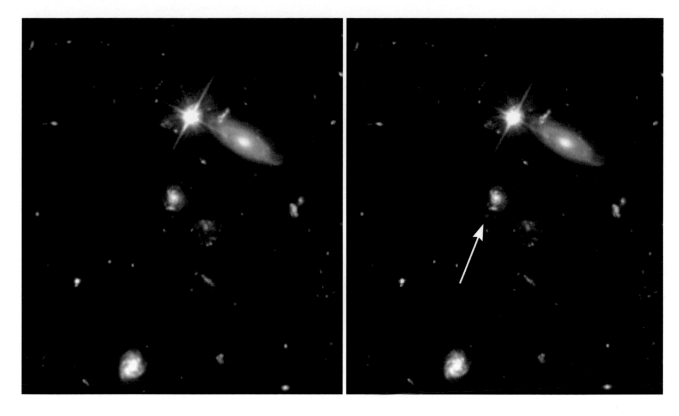

excess of one. This was achieved by 'piggy-backing' on a major programme, called the Great Observatories Origins Deep Survey (GOODS), which had been allocated a large amount of observing time to use the Hubble Telescope's ACS to obtain images of remote galaxies at a range of different wavebands (colours). Because the instrument was sufficiently sensitive to detect objects that were several times fainter than the expected peak brightness of supernovae with redshifts in the

region of 1.0 to 1.6, there was good reason to suppose that the search would find at least a few remote supernovae in this critical redshift range.

The search was so successful that by June 2004 Riess was able to announce, in *Astrophysical Journal*, the discovery of 16 new Type Ia supernovae, six of which had redshifts in excess of 1.25 (these were six of the seven highest-redshift supernovae known at that time), the highest having a redshift of 1.6. By

Distant supernova 2002dd shows up in the right hand image as the red spot identified by the arrow. The left hand image shows the same region of sky 7 years before the event. Supernova 2002dd and its host galaxy are seen as they were when the universe was about half of its present size.

These are images of three of the most distant supernovae known, discovered by the Hubble Higher-Z Supernova Search team. The lower images show the supernovae (identified by arrows) together with their host galaxies; the upper images show the host galaxies before the supernova explosions.

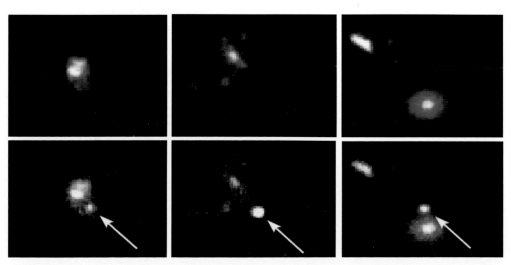

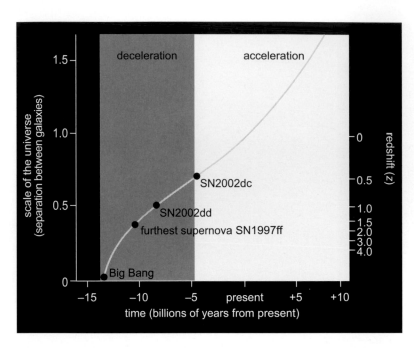

years ago, the universe switched over from deceleration to acceleration. Whereas the gravitational attraction of matter was dominant when the universe was smaller and denser, its influence weakened as matter became more dilutely spread through the expanding volume of space. Nowadays, the universe is dominated by dark energy, and is expanding at an accelerating rate.

What is driving the acceleration? Dark energy appears to be the culprit, but what is the nature of this mysterious entity? To find one possible candidate and mechanism, we need to look back nearly 90 years to explore what has come to be known as 'Einstein's greatest blunder'.

Supernova data indicates that the expansion of the universe switched from slowing down to speeding up just over 5 billion years ago. The points indicate the redshifts of some of the particular supernovae mentioned in the text.

plotting peak brightness against redshift, they were able 'robustly' to extend the Hubble diagram into the region between $z = 1$ and $z = 1.6$. The results showed clearly that these remote supernovae are distinctly brighter than would be the case if grey dust absorption or the evolution of supernova properties were markedly affecting their observed brightness (at least for many of the simpler models of these effects). By combining their high-redshift data with results for around 170 lower-redshift supernovae, they found compelling evidence (with better than 99 percent statistical confidence) for recent acceleration and past deceleration. Their results indicated that the turnover from deceleration to acceleration occurred at a redshift of around 0.46 (their published figure was 0.46 ± 0.13), in which case the switch to acceleration must have taken place about 5 billion years ago, when the universe was about two-thirds of its present size.

Type Ia supernovae seem to be telling a clear and consistent story. The universe contains matter and dark energy, with the mean density of matter being about 30 percent of the critical density and the dark energy density about 70 percent of critical. The acceleration seems to be real. During the first 8 billion years or so of the history of the universe, its expansion was slowing down. Then, about 5 or 6 billion

CHAPTER TEN

Einstein's Greatest Blunder?

That the universe might contain an extra ingredient which could cause it to expand at an accelerating rate is a concept that grew out of Albert Einstein's attempts to resolve a particular cosmological conundrum which he faced, in 1917, when he set out to apply his newly formulated general theory of relativity to the universe as a whole. At that time, many, though by no means all, astronomers believed that we inhabit an 'island universe', a great system of stars surrounded by a sea of emptiness, and the fact that our Galaxy is merely one of a vast multitude of similar systems had not been proven. The recession of the galaxies and the expansion of the universe had yet to be discovered, and it was generally believed that the universe was static (neither expanding nor contracting).

To enable his theory to accommodate a static universe, Einstein added an extra mathematical term to his equations. This extra component, which was called the cosmological term, incorporated a constant – known as the cosmological constant and denoted then by the lower case Greek letter lambda (λ), but nowadays by the upper case Λ – which modified the curvature of space. With the additional term in place, the curvature of space would depend not only on the distribution of mass and energy, but also on the value of the cosmological constant. If the cosmological constant were positive, the extra term would manifest itself as a form of 'cosmic repulsion' acting in the opposite direction to gravity; conversely, a negative cosmological constant would provide an additional attractive effect. Over astronomically small distances, and in regions where the density of matter was substantial, cosmic repulsion would be negligible. Its effects would be imperceptible, for example, in the motion of the planets of our Solar System. But at sufficiently large distances, over vast regions of space where the average density of matter is very low, its effect could be very significant. If the cosmological constant had a particular critical value, then over the distances that separate the stars (or galaxies, as we would now say) cosmic repulsion would exactly counterbalance the attractive force of gravity, and the universe would be static.

In effect, Einstein had adjusted gravitation to fit the then contemporary paradigm of a static universe. His static universe contained a finite density of matter and was endowed with a positive value of the cosmological constant; it had positive curvature (analogous to the curvature of a sphere), was finite in extent, and was maintained in its static state by the finely tuned balance of the opposing influences of gravitational attraction and cosmic repulsion.

In that same year, 1917, Dutch astronomer Willem de Sitter showed that with a cosmological constant present, an empty universe (one which contained no matter) would expand at an accelerating rate. Space would continue to stretch – like an elastic sheet – at a greater and greater rate as time went by. If small test particles were placed in that space, they too would begin to accelerate away from each other. The concept of a universe with no matter bears no resemblance to present-day reality, of course, and the de Sitter model did not appeal to Einstein at the time (although, ironically, it has relevance to some modern ideas about what may happen to the universe in the very distant future – see Chapter 11).

Some five years later, the Russian meteorologist and mathematician Alexandr Friedmann derived equations, based on general relativity, which showed that the universe could expand forever, or expand to a finite size and then collapse, even without the 'extra' cosmological term. But because Friedmann was working in relative isolation, and the expansion of the universe had yet to be established by observations, the idea did not gain much immediate currency and passed largely unnoticed at the time. So, although the theoretical possibility of an expanding universe had been demonstrated in the early 1920s by de Sitter and – in more realistic fashion – by Friedmann, it fell to observational astronomers to discover, and to convince the astronomical community, that the universe is indeed expanding. The pivotal step in that process was Edwin Hubble's discovery of the relationship between the redshifts and distances of the galaxies, and his publication in 1929 of what has come to be known as the Hubble law (see Chapter 2).

Faced with observational evidence for an expanding universe that was dynamic and changing, rather than static, Einstein abandoned the cosmological term (and with it, the cosmological constant) in 1931, stating then that the theory of general relativity could account for the expansion in a natural way without a 'lambda term.' In 1932, Einstein and de Sitter developed a simpler model of the universe, with zero lambda, flat-space, and only the gravitational influence of matter to consider. The 'Einstein–de Sitter' model was one in which the kinetic energy of the receding galaxies was exactly equal and opposite to the energy of gravitational attraction – a flat-space universe in which the rate of expansion would slow down relentlessly, but not become zero until the infinite future. This type of universe, which has the critical mean density that enables it to 'sit on the fence' between the open (ever-expanding) and closed (ultimately collapsing) Friedmann models, was to find resonance among theoreticians in the latter part of the twentieth century.

Looking back on it many years later, Einstein defended both his introduction of the cosmo-

logical term and his subsequent rejection of it, pointing out that while the extra mathematical term was entirely consistent with relativity, it was not necessary. He also, perhaps somewhat ruefully, pointed out that he would never have added it to his equations had Hubble's discovery of the expansion of the universe been made at the time when he (Einstein) was developing his general theory of relativity.

Writing in 1970, some 15 years after Einstein's death, renowned astrophysicist George Gamow asserted that Einstein, in conversation with him, had said that the introduction of the cosmological term had been the biggest blunder he ever made in his life. One reason he may have made that remark is that, by introducing the extra term to make the universe static, he missed out on the chance of predicting the expansion of the universe directly from his own equations well in advance of its discovery by observational astronomers.

However, although Einstein himself abandoned the cosmological constant, there were others who continued to treat the concept seriously, most notably, English astrophysicist Sir Arthur Stanley Eddington (the greatest contemporary English exponent of Einstein's theories of relativity, and the astronomer who made the crucial measurements of the bending of light at the 1919 eclipse which provided dramatic confirmation of general relativity), and the Belgian priest and distinguished physicist, Abbé Georges Lemaître. Eddington had suggested as early as 1924 that the galaxy redshifts observed by Vesto Slipher (see Chapter 2) might be a manifestation of the repulsive effect of the cosmological constant, and in 1930, pointed out a disturbing feature of the static universe that Einstein himself does not appear to have appreciated at the time when he devised that model, namely that it is inherently unstable. A tiny change in the overall density of matter, or in the way in which matter was distributed, would be enough to upset the balance and cause the whole of the universe – or localised regions within it – to expand or contract. For example, if the stars (galaxies) were nudged apart by a small amount, cosmic repulsion would gain the upper hand, causing them to fly apart at an accelerating rate, whereas if they were nudged a little closer, gravity would win the battle and the stars (galaxies) would begin to fall together. This realisation spurred Eddington to investigate expanding solutions of Einstein's equations which included a positive cosmological constant.

However, three years earlier (in 1927) Lemâitre, who was aware of Slipher's observations and Hubble's ongoing programme, had already independently derived detailed sets of equations, based on general relativity with the lambda term included, which described expanding models of the universe. These turned out to be essentially the same as those which had been derived earlier by Friedmann, but of which Lemâitre had been unaware at the time.

Whereas Lemâitre was comfortable with the idea of a universe that originated at a particular instant by expanding from a singularity (a highly-compressed initial state), Eddington was not. Building on Lemâitre's work, Eddington developed a model universe (the Eddington–Lemâitre model) which remained in the finely-balanced Einstein state for an indefinitely long period of time until something tipped the balance in favour of the cosmological constant, whereupon the universe began to expand at an accelerating rate. Lemâitre, on the other hand, found an initial beginning to the universe perfectly acceptable, and proposed that all the matter in the universe had emerged from a massive 'primeval atom' which – by analogy with what he knew at that time about radioactive decay – had disintegrated, scattering matter forth (Lemâitre's idea was the forerunner of what later came to be called the Big Bang). He suggested that the universe expanded thereafter at a decelerating rate until it approached the Einstein static state, at which stage the expansion virtually ceased. The universe then loitered in a quasi-static state until such time as the very slow ongoing expansion tipped the balance in favour of cosmic repulsion, and the universe entered a phase of accelerating expansion akin to that proposed by de Sitter for an 'empty' universe – an ongoing perpetually accelerating expansion into the indefinite future.

The age problem

The Eddington–Lemâitre model and the Lemâitre 'hesitation' model each provided a very neat solution to the 'age crisis' that was brewing in the 1930s. At that time, Hubble's measured value of the Hubble constant was 530 km/s/Mpc – seven or eight times greater than present estimates – and the resulting value for the Hubble time (the time to expand to its present size at a constant rate of expansion), which was the reciprocal of the Hubble constant, was disturbingly low – about 1.85 billion years (even less – about 1.2 billion years – in the case of an Einstein–de Sitter critical density universe). This was less than contemporary estimates of the age of the Earth, derived from the radioactive decay of elements in terrestrial rocks. Surely the Earth could not be older than the universe within which it was contained? To make matters much worse, estimates of the lifetimes of stars, based on the erroneous assumption that in their lifetimes stars would convert all of their mass (rather than a small fraction of it) into energy, suggested that the ages of stars were vastly greater than the Hubble time. Even when Hans Bethe worked out, in 1939, the basic framework of the currently-accepted theory of the nuclear reactions that power the Sun, and realised that its total lifetime would be in the region of 10

Whereas the age of a decelerating universe, such as the Einstein-de Sitter model, is less than the Hubble time, the age of a Lemâitre or Eddington-Lemâitre universe can be arbitrarily greater.

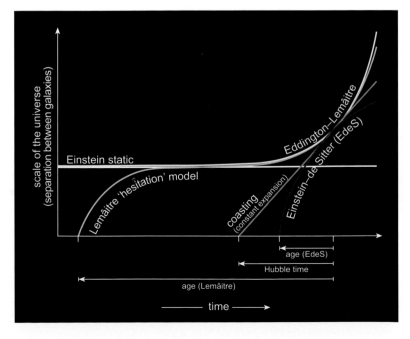

billion years, the ages of the stars still seemed to be about five times greater than the expansion age of the universe.

If Lemâitre's hesitation model, or the Eddington–Lemâitre model, were valid, the age derived from the Hubble constant would be incorrect because it uses the present-day value of the (accelerating) expansion rate to calculate how long it has taken for galaxies to recede to their present distances. By adjusting the duration of the long hesitation period between the decelerating and accelerating phases of expansion, or the duration of the initial Einstein static state, the overall age of the universe could be made as long as one wished. So, by incorporating a positive cosmological constant, both of these models could circumvent the age problem.

The 'age problem' was still an issue in the late 1940s, and was one of the factors that prompted Thomas Gold and Hermann Bondi, together with Sir Fred Hoyle, to develop the Steady State alternative to what is now called the Big Bang. Proponents of the Steady State theory contended that, despite the fact that the galaxies are receding from each other, the universe on the large scale is homogeneous (the same everywhere) and isotropic (looks the same in every direction) at all times (this set of conditions constituted the so-called perfect cosmological principle). Thus, a large volume of space today would contain the same number of galaxies as would an equal-sized volume of space billions of years ago or billions of years hence. In order to satisfy the perfect cosmological principle, galaxies, as they moved apart, had to be replaced by new ones formed out of matter that was continuously created in space. According to the Steady State theory, space and time were infinite – the universe had no 'beginning', nor would it have 'an end'. The Steady State theory had a timeless quality that appealed to many people (there was no need to tackle questions such as, 'Where did the matter for the Big Bang come from?' or 'What happened before the Big Bang?' – the universe had always been as it is and always would be). To make the theory work, galaxies had to move apart at an accelerating rate; Steady State theory required a cosmological constant in a different guise.

But part of the justification for the Steady State theory had been undermined in the 1950s by accumulating observational evidence that raised doubts about the age problem. The first major step came in 1952 when Walter Baade, using the 200-inch (5-metre) telescope at Mount Palomar, and A D Thackeray, using the 74-inch Radcliffe telescope in South Africa, discovered that there are two different types of Cepheid variable, Type I and Type II, of which the Type I variety are inherently about two to three times more luminous than the Type II. Hubble had based his distance scale on studies of Cepheid variables in the Magellanic Clouds which had been carried out in 1912 by Harvard astronomer, Henrietta Leavitt, and had assumed (for no one knew otherwise at the time) that the Cepheids he had identified in the Andromeda galaxy were of the same sort. Baade realised that the Cepheids which Leavitt had studied were the less luminous Type II specimens, whereas the ones Hubble had detected in other galaxies were the more luminous Type I variety. In order for Cepheids to appear as faint as they did, the galaxies within which they were embedded had to be considerably further away than had previously been thought. Consequently, the Hubble constant would have a smaller value, and the Hubble time would be longer. Baade's results immediately reduced the value of the Hubble constant to about half Hubble's original figure, and

According to the Steady State theory, as galaxies A, B, C and D recede from us and from each other, they will be replaced by new galaxies (E, F, G, and K). Consequently, the average number of galaxies in a given box of space will remain the same at all times.

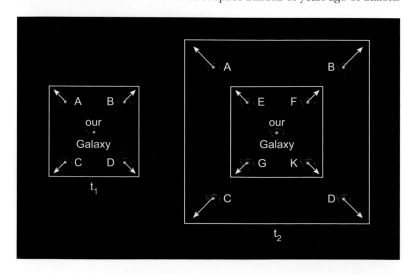

doubled the size and the age of the universe.

American astronomer, Allan Sandage – who had begun his career as an assistant to Hubble – then embarked on an extensive programme of observations to measure galaxy distances, with the aim of refining estimates of the Hubble constant and with the ultimate goal of deciding between the various competing models of the evolution of the universe (a programme with which, half a century on, he is still actively involved). By 1956, his data had reduced the value of the Hubble constant to somewhere in the region of 75 km/s/Mpc, which corresponded to a Hubble time of around 13 billion years – comfortably larger than the age of the Sun. It began to look as if the age problem might no longer be an issue, and the case for a positive cosmological constant weakened as a result.

Although the Steady State theory continued to provoke lively discussion and debate – and to stimulate observational cosmologists to improve their efforts to measure redshifts and distances – for nearly two decades, it was eventually sidelined by the growing body of evidence in favour of Big Bang cosmology. A significant piece of evidence was the discovery, in 1965, of the microwave background radiation which showed that the universe had been in a hot dense state in the past – something which could not have happened in a constant-density Steady State universe. Once again, a possible raison d'être for the cosmological constant had faded away.

Nevertheless, for a short time during the 1960s, cosmologists found another possible reason for revisiting the cosmological constant. As technological advances enabled them to peer further and deeper into space, they were able to detect extremely luminous objects called quasars at higher and higher redshifts. They found that when they plotted the observed number of quasars against redshift, there seemed to be a well-defined peak in their numbers at a redshift in the region of $z = 2$. If the universe had loitered for a considerable time at about one-third of its present size (i.e. around $z = 2$) during the transition from deceleration to acceleration in a Lemâitre-type

hesitation universe, then this plateau in the expansion history of the cosmos would have spanned a range of distances and look-back times. Consequently, astronomers would see more quasars with this particular value of redshift than with smaller or larger redshifts. Although these results raised the spectre of a hesitation universe endowed with a positive cosmological constant, as they found out more about quasars, astronomers soon came up with a convincing alternative.

Quasars are believed to be extreme examples of active galactic nuclei (an active galactic nucleus is a compact, exceptionally luminous and often variable energy source lurking at the heart of an active galaxy, which is believed to be powered by accretion of matter onto a supermassive black hole). Nowadays, their prevalence at redshifts of two to three is usually taken to indicate that the evolution of violent activity in quasars reached a peak when the universe was between a third and a quarter of its present size, which corresponds to looking out to redshifts of between 2 and 3; we see more quasars at those redshifts simply because there was more quasar activity then than at any other time. Having briefly popped its head above the parapet, the cosmological constant had to duck down and take cover again.

With the apparent sidelining of the age problem, the demise of the Steady State, and the resolution of the quasar redshift plateau issue, the cosmological constant became a 'no-go area' for many cosmologists. In the absence of clear evidence for its existence, most cosmologists simply assumed that the value of lambda was zero. But, although it languished in relative obscurity until the last decade of the twentieth century, it was a concept that would not lie down or go away completely.

The cosmological constant made a brief foray into the collective astronomical consciousness again around 1975 when Caltech astronomers, James E Gunn and John B Oke, made a series of measurements of the relationship between the redshift and apparent brightness of giant elliptical galaxies – highly-luminous objects that could be seen out to very great distances. From their observational

The Hubble image on the left shows the brilliant quasar, 3C273, to be star-like in appearance (the spikes are an instrumental effect). The image on the right, taken with the Hubble telescope's ACS coronagraph, which blocks out the dazzling core of the quasar, shows details of the host galaxy within which it is embedded.

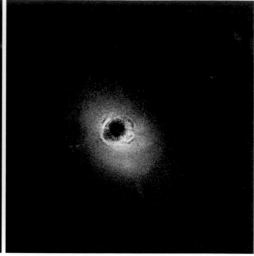

Practically all of the stars in a young cluster lie on the main sequence, but if the cluster is as old as the Sun, stars which are more massive than the Sun will have left the sequence. If older still, only the bottom end of its main sequence will remain. The 'turn-off point' gives the age of the cluster.

data, they produced a Hubble diagram that appeared to curve away from the straight-line Hubble law in a way that, according to Gunn and University of Texas astronomer Beatrice M Tinsley, could be due to the effect of a finite cosmological constant. The validity of the results hinged on whether or not giant elliptical galaxies formed a sufficiently homogeneous set of objects to act as a valid standard candle,

but unfortunately this turned out not to be the case (it became apparent that giant ellipticals have a large spread of sizes and luminosities, and that their luminosities have changed over time). Once again, the cosmological constant was put back in its box.

But the age problem continued to haunt, tease and tantalise astronomers. Over the decades, the measurement of galaxy distances, and the derivation of an accurate value for the Hubble constant, proved to be a phenomenally challenging task, largely because of the difficulty of finding sufficiently reliable standard candles (or standard rulers) and the equally intractable problem of how to take account of evolutionary changes in distance indicators over cosmic timescales. During the 1960s and 1970s, estimates of the Hubble constant (H_0) derived by different observers using a range of techniques were generally in the region of 50–100 km/s/Mpc, which gave Hubble times ranging from about 20 billion years (for $H_0 = 50$ km/s/Mpc) to about 10 billion years (for $H_0 = 100$ km/s/Mpc), and corresponding flat-space Einstein–de Sitter ages from around 13 billion years to about 6.7 billion years respectively. By the 1970s, estimates for the ages of the oldest stars were coming in at around 15–18 billion years. Whereas those stellar ages fitted reasonably comfortably into the timescale of a relatively low-density open universe if the value of the Hubble constant were near to the lower end of the range (around 50 km/s/Mpc),

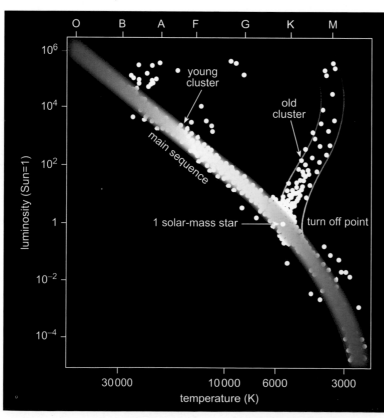

The Pleiades star cluster is a youthful one which is only about 100 million years old. It contains more than 500 stars (six of which are readily visible to the unaided eye under good conditions), and lies at a distance of about 380 light-years.

they appeared to be in conflict with values at the higher end.

Refining the ages of stars

Calculating reliable ages for the oldest stars is an extremely difficult task. One clue comes from looking at their spectra and measuring the proportion of heavier elements that they contain (by quaint convention astronomers refer to anything heavier than hydrogen and helium as a 'metal'). The first generations of stars formed from a near pristine mix of hydrogen and helium. Most of the heavier elements found in later generations of stars, such as the Sun, were produced subsequently by nuclear reactions that took place inside stars and scattered forth into the interstellar clouds from which Sun-like stars subsequently formed, by supernova explosions and stellar winds (outflows of gas from the atmospheres of elderly stars). Consequently, very old stars from the earliest generations are 'metal poor'.

Globular clusters – massive near-spherical clusters that contain from a few tens of thousands to around a million stars, and which are found in the halo of our Galaxy (and in the haloes of other galaxies, too) – are among the oldest objects in the universe. One way of calculating the age of a globular cluster is to plot its constituent stars as points on a colour-magnitude, or Hertzsprung–Russell, diagram (see Chapter 1) in order to identify its main sequence. As we saw in Chapter 1, the length

of time for which a star remains on the main sequence depends on its mass: the more massive the star, the hotter and more luminous it is, and the faster it uses up its reserves of fuel. When a star has consumed all the available hydrogen fuel in its core, it moves off the main sequence and eventually becomes a red giant. In a young, recently-formed cluster (such as the Pleiades), most of the stars will still be on the main sequence, but in an old, evolved cluster, the only stars still on the main sequence will be the cooler low-mass ones at its lower end. The 'turn-off point' on the cluster's main sequence corresponds to stars which are on the point of leaving the sequence (having virtually exhausted all the hydrogen fuel in their cores), the ages of which can be calculated from the theory of stellar evolution. Because the stars which populate a globular cluster are all believed to have formed at much the same time, during a rapid bout of star formation early in the history of the Galaxy, the age of the cluster will be the same as the ages of the stars at the turn-off point.

A second approach relies upon identifying white dwarfs (the shrunken remnants of stars which are no longer generating energy by means of nuclear reactions) in globular clusters – a difficult task because such stars are exceedingly faint. These stars gradually cool and fade. Knowing the rate at which white dwarfs cool down and fade, astronomers can calculate the age of a cluster by identifying the faintest,

(Left) The globular cluster M4 contains more than 100,000 stars and, at a distance of some 7,000 light-years, is the nearest cluster of its kind. (Right) This Hubble Space Telescope image shows a close-up view of a small region (0.63 light-years across) within M4 and reveals seven white dwarf stars (inside the blue circles).

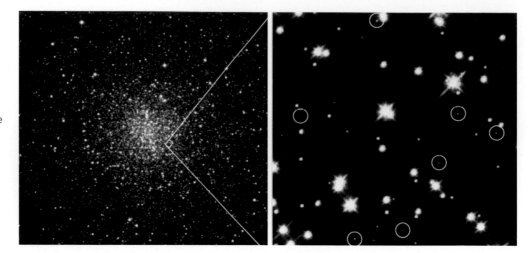

coolest white dwarfs it contains.

A third line of attack uses a technique similar to that which is used by geologists to work out the ages of rocks – radioactive dating. At the time when a star was born, the mix of materials from which it formed would include radioactive elements, such as thorium or uranium, which have very long half-lives (half life is the time taken for half of the atomic nuclei in a sample of material to decay). If astronomers can identify lines due to elements such as these in the spectra of old, metal-poor stars (a very challenging task), they can determine the present ratio of the amount of each radioactive element relative to the amounts of stable elements in the atmosphere of a star. By comparing those ratios with theoretically computed ratios for the relative proportions of these elements at the time when they were formed, astronomers can work out how much time has elapsed since the star was born (more reliable results can be obtained by comparing the observed relative proportions of two or more different types of radioactive elements, taking into account the fact that each decays at a different rate).

The emergence of inflationary cosmology in the 1980s, and its enthusiastic adoption by theoreticians, brought the simmering age problem back to centre stage. As we have seen (Chapter 8) inflationary cosmology requires the geometry of space to be flat (or indistinguishably close to flat) and the mean density of the universe to be indistinguishably close

to the critical density (i.e., $\Omega = 1$). After its short-lived but dramatic initial bout of accelerating expansion, the ongoing evolution of the inflationary universe, in its simplest guise, behaves just like the flat-space Einstein–de Sitter model: gently decelerating ongoing expansion, with an expansion age equal to two-thirds of the Hubble time. By the mid-1990s, most (but not all) estimates of the Hubble constant were converging on values in the region of 65–80 km/s/Mpc, yielding Hubble times of around 12–15 billion years and ages in the region of 8–10 billion years for flat-space, critical density models. At that time, estimates of the ages of the oldest globular clusters were coming out in the region of 16 billion years ($\pm$ 3 billion years). Given that the best part of a billion years probably had to elapse between the origin of the universe and the formation of the first globular clusters, the age problem seemed to be rather acute – especially for the favoured flat-space, critical density model.

The inflationary model was also running into a density problem. As we have seen in previous chapters, Big Bang nucleosynthesis arguments implied that baryonic matter could contribute only 4–5 percent of the critical density. To preserve their neat, flat-space, critical density universe, cosmologists had to assume that enough non-baryonic dark matter would eventually be identified to ensure that the overall density was indeed equal to the critical density. Thus a consequential implication of the inflationary hypothesis was that it required the existence of large amounts of non-baryonic matter. Through the 1980s and into the 1990s, there was a general expectation that the universe would turn out to be a critical density flat-space one with 95 percent of the total density being contributed by cold dark matter. Yet as more and more observational evidence accumulated, it became more and more obvious that this was not so, and that the total density of matter in all its forms (luminous and dark, baryonic and non-baryonic) was actually no more than 30 percent of critical ($\Omega_M = 0.3$).

The favoured theoretical scenario in the 1980s and early 1990s was the 'simple', or 'standard' cold dark matter model (designated SCDM), with flat-space, and a combination of baryons (about 5 percent) and cold dark matter (about 95 percent) making up a total matter density equal to the critical density ($\Omega_M = 1$). It was a successful theory in that supercomputer simulations of the formation and evolution of structures within an SCDM universe matched more closely with the real universe than other alternatives such as the hot dark matter model (which required most of the mass-energy to be bound up in fast-moving neutrinos). But there was a problem. If current theory is valid, the seeds for the formation of structure in the universe were the marginally denser patches (density fluctuations) that were generated by inflation and which are imprinted on the CMB as a distribution of warmer and cooler patches (temperature fluctuations). Gravity then pulls these overdense regions together to form galaxies, clusters and larger-scale structures. After the COBE satellite (see Chapter 8) had revealed these fluctuations for the first time, in 1992, theoreticians were able to run SCDM simulations based on the observed temperature fluctuations. The results were discouraging. SCDM predicted much more structure (and about ten times as many galaxy clusters) than was actually observed in the real universe. The COBE results appeared to rule out the simple CDM inflationary model.

Theoreticians tried a few 'fixes', such as adding to the mix of baryons and cold dark matter a component of hot dark matter in the form of neutrinos (vCDM), extra radiation due to the annihilation of particles such as tau neutrinos (τCDM), or tinkering with inflation theory's predictions about the properties of primordial density fluctuations (this was called the 'tilted' cold dark matter model – TCDM); but problems remained.

By 1995 it was becoming clear to a number of cosmologists that if space is indeed flat, and the total density of matter (luminous and dark, baryonic and non-baryonic) is only about one-third of critical, then the inflationary flat-space model (which has produced so many benefits) can only be valid if around two-thirds of the mass-energy in the universe exists in some other form. In that year, Lawrence M Krauss of Case Western University, Cleveland, and Michael S Turner, of the University of Chicago and Fermilab wrote a paper entitled 'The Cosmological Constant is Back' in which they pointed out that problems relating to the age, overall density and curvature of the universe, and issues relating to the way in which large-scale structure formed, could all be overcome if two-thirds of its total mass-energy consisted of some smoothly-distributed form of invisible energy – dark energy – such as the cosmological constant. The cosmological constant (or some other form of repulsive dark energy) would stretch the age of the universe beyond the Hubble time, so resolving the age problem. Based on the observational data available at the time, their 'best fit' model was of a universe with $H_0 = 70–80$ km/s/Mpc, $\Omega_M = 0.3–0.4$, and $\Omega_\Lambda = 0.6–0.7$ (where the symbol Ω_Λ denotes the dark energy density parameter – the density of dark energy expressed as a fraction of the critical density).

Almost simultaneously, Jeremy P Ostriker of Princeton University, and Paul J Steinhardt of the University of Pennsylvania, published a paper entitled 'Cosmic Concordance'. They showed that the constraints imposed on the overall properties of the universe by the available measurements (at that time) of the angular size of the fluctuations in the CMB, large-scale structure, the Hubble constant, the overall matter density and the ages of the stars, are all consistent with a spatially flat model incorporating a cosmological constant with a density parameter in the region of $\Omega_\Lambda = 0.65 \pm 0.1$.

By 1997, Turner was able to assert that, whereas the best available data from the CMB, matter clustering, mean density and Hubble constant had ruled out SCDM and seriously disadvantaged other options such as νCDM and τCDM, 'the only CDM model consistent with all present observations is ΛCDM' (baryons plus cold dark matter plus a dominant component of dark energy in the form of the cosmological constant).[1] Another option, OCDM (an open universe with baryons and CDM totalling about one-third of critical density, but with curvature and zero dark energy) remained a possibility, but was utter anathema to inflationists! When cosmologists began to investigate models with cold dark matter and a positive cosmological constant, they found they could simulate the development of structure in the universe in a way that matched rather well with the growing body of observational data relating to the distribution and clustering of matter in the universe at different times in its history. ΛCDM became the new paradigm of the late 1990s.

All of these ideas had emerged before the epoch-making supernova results of 1998, which showed that the universe is indeed accelerating; before the BOOMERanG and MAXIMA microwave background results of 2000, which confirmed the flatness of space; and before the subsequent detailed WMAP, 2dFGRS and SDSS results which have so greatly enhanced the precision with which the key cosmological parameters have been measured. When coupled together, the supernova results, the spectrum of microwave background temperature fluctuations, the 2dFGRS and SDSS studies of galaxy distributions, Lyman-alpha cloud measurements and weak gravitational lensing data all point towards the same concordance cosmology – in round figures a flat-space universe with a matter density of about 30 percent of critical ($\Omega_M \approx 0.3$, comprising $\Omega_b \approx 0.05$ and $\Omega_{CDM} \approx 0.25$) and a dark energy density of about 70 percent of critical ($\Omega_\Lambda \approx 0.70$).

Whereas the discovery of the accelerating expansion of the universe in 1998 came as a great surprise to many astronomers, to infla-

tionary cosmologists it came as no surprise at all. Rather, it was seen as the anticipated final piece in a cosmic jigsaw that fitted together the low observed mean matter density ($\Omega_M = 0.2$–0.3), the flat-space geometry which was hinted at by the CMB data available at that time, the observed properties of the CMB fluctuations as revealed by COBE and ground-based measurements (both of which matched the predictions of inflationary theory), and the measured value of the Hubble constant. ΛCDM fitted the bill.

Since the start of the current millennium, determinations of the ages of the oldest stars – from globular cluster main sequence turn-off, the cooling of white dwarfs, and radioactive dating – have all (within their individual spread of errors) converged on values in the range 12–13 billion years. The age of the universe for an $\Omega_M = 0.3$, $\Omega_\Lambda = 0.7$ universe with H_0 in the region of 70 km/s/Mpc works out at around 14 billion years, which neatly accommodates stellar ages in the region of 12–13 billion years. By contrast, the OCDM model is ruled out by the flat geometry of space, its failure to match the measured spectrum of microwave background temperature fluctuations, and ongoing difficulties with the age problem (an OCDM model with $\Omega_M = 0.3$ has an expansion age of about 11.3 billion years – less than the ages of the stars).

Dark energy of some kind is required by the observations. The cosmological constant is a potential candidate that has been around for nearly 90 years. After languishing in relative obscurity for many decades, despite popping its head above the parapet from time to time, the cosmological constant is now back in the frame with a vengeance. Is Einstein's much-maligned constant, disowned by its instigator, the key to the overall balance of matter and energy in the universe and the driver of cosmic acceleration? Could his 'greatest blunder' turn out instead to have been inspired insight into the workings of the cosmos?

CHAPTER ELEVEN

Dark Energy - The Prime Mover

Dark energy is the extra ingredient that cosmologists require in order to balance the density of the universe and account for why its expansion is accelerating. Yet the very term 'dark energy' – evocative but at the same time nebulous – reflects our utter ignorance of its true nature.

Unlike matter, dark energy does not clump together under the influence of gravity. If it did, it would affect the internal dynamics and motions of galaxies and clusters, and would reveal itself in phenomena such as gravitational lensing and the evolution of cosmic structure. Instead, it seems to be smoothly distributed through the whole of space – affecting the global properties and expansion history of the universe as a whole without apparently affecting, or itself being influenced by, localised blobs of matter such as galaxies and clusters.

Although it makes up more than 70 percent of the mass-energy of the universe today, its repulsive influence must have been insignificant compared to the gravitational attraction of matter when the universe was younger, otherwise it would have prevented matter from falling together to make stars, galaxies, and clusters. Because the density of matter is inversely proportional to the cube of the expansion factor of the universe, if we look backwards in time, the matter density increases rapidly (increasing by a factor of eight each time the scale of the universe shrinks to half its previous value). In order for dark energy to be the major component of the cosmic mass-energy budget now, but an insignificant one in the past, its density must either remain constant or change much more slowly than the density of matter. If that is the case, then if we were to run the cosmic clock backwards, the dark energy density would either remain the same or would increase much more slowly than the matter density, and would quickly become inconsequential by comparison. Conversely, despite starting out with a very low density relative to that of matter, dark energy's much gentler decline would ensure that it eventually became the dominant component.

Essential attributes

Dark energy is a smoothly distributed, slowly declining (or constant) commodity that has two key roles to fulfil – it has to make up the shortfall in density between the critical density and the actual density of matter and radiation, and it has to provide the repulsive influence that is driving the accelerating expansion of the universe. To meet these twin demands, it needs to possess two essential attributes – positive energy density and negative pressure.

According to Einstein's theory of relativity, energy and mass are equivalent, and matter, radiation and other forms of energy all exert a gravitational influence. To make up the measured shortfall, dark energy has to make a positive contribution to the overall mass-energy density of the universe equal to about 70 percent of the critical density – that seems straightforward enough. But, at first glance, the concept of negative pressure seems alien to everyday experience. When we think of pressure, we tend to associate the idea with something that pushes outwards, like the air inside an inflated balloon. As we pump more and more air into a balloon, the pressure exerted by the gas molecules rushing around

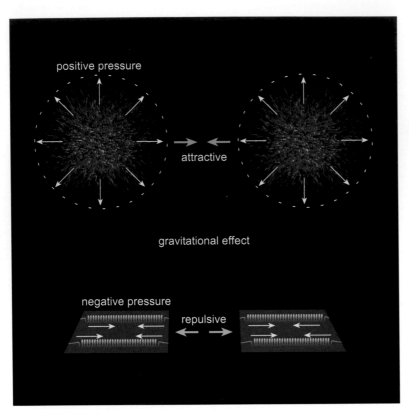

positive pressure

attractive

gravitational effect

negative pressure

repulsive

Positive pressure, which pushes outwards (as in the two gas-filled balloons shown here) gives rise to an attractive form of gravity whereas negative pressure (represented by the stretched springs shown here) gives rise to a repulsive form of gravity.

inside it and bouncing off its inside skin, provides an outward push that causes the balloon to expand. If we increase the pressure too much – by pumping in more air, or by heating the balloon (the hotter the gas, the faster its molecules move, and the greater the pressure they exert) – the balloon eventually bursts. Positive pressure pushes outwards like a compressed spring. Negative pressure, by contrast, pulls together, like a stretched spring or an elastic sheet.

According to Einstein's equations, pressure itself is a source of gravitation. But whereas positive pressure, like matter, gives rise to attractive gravity which pulls things together, negative pressure gives rise to a repulsive form of gravitation that pushes things apart. If the universe contains a negative pressure component (dark energy) the dynamics of its expansion will be determined by the competition between the attractive influence of matter, radiation and other forms of energy (such as kinetic energy), and the repulsive influence of dark energy. If the repulsive influence exceeds the attractive one, cosmic repulsion

will gain the upper hand, and the expansion of the universe will accelerate.

General relativity specifies that the rate of change of cosmic expansion depends on the overall mean density of matter and energy (symbol, ρ) plus three times the net overall pressure (P). If the sum of these quantities ($\rho + 3P$) is positive (greater than zero), the expansion will decelerate, but if the sum of these quantities is negative (less than zero), then the expansion will accelerate. The pressure exerted by the dilute and thinly spread distribution of matter and radiation in space is usually considered to be negligible, so that – in the absence of dark energy – the density of matter and radiation controls the dynamics of the universe. However, if negative-pressure dark energy is present, the magnitude of its negative pressure need only be greater than one-third of the overall mass-energy density of the universe in order for ($\rho + 3P$) to be negative, in which case the expansion will accelerate. The bottom line for dark energy is that its pressure must be sufficiently negative to overcome the gravitational attraction of all the mass-energy density in the universe.

The ratio of pressure to density (P/ρ) is called the equation of state, and is usually denoted by the symbol w (i.e. $w = P/\rho$). Dark energy on its own will have a repulsive influence if this ratio is more strongly negative than $-1/3$ (i.e. $w < -1/3$), but because the total mass-energy density includes both matter (the density of which is equivalent to slightly less than half the dark energy density) and dark energy, the ratio of dark energy pressure to dark energy density (the equation of state of dark energy) has to be more strongly negative ($w < -1/2$) if repulsion is to exceed attraction at this particular time in the history of the universe. Could the cosmological constant meet this criterion? By definition, its pressure and density are equal and opposite ($P = -\rho$), so the ratio of pressure to density is constant at all times and has the value -1 ($w = -1$). The pressure associated with the hypothesised cosmological constant is certainly sufficiently negative to be driving accelerated expansion today.

Lambda redefined

The cosmological constant is a prime dark energy suspect. But (assuming that it exists), what is its nature? When Einstein added this constant to his equations, he regarded it as a property of space – an extra ingredient, which modified the geometry of space-time. Modern cosmologists and particle physicists look at the cosmological constant in a very different, although mathematically equivalent, way, treating it as an extra form of energy – called vacuum energy – which adds a constant energy density to the universe.

The existence of vacuum energy is a logical consequence of the Uncertainty Principle of quantum mechanics. As we saw in Chapter 1, one of the implications of this principle is that we cannot know the precise amount of energy contained within a microscopic physical system (or within a miniscule 'box' of apparently empty space) over a very short interval of time. Consequently, over short enough intervals of time, energy levels within an apparently empty vacuum can fluctuate. These 'vacuum fluctuations' can provide enough energy to make pairs of particles and antiparticles (particle–antiparticle pairs) but, provided that the particles and antiparticles annihilate each other within the interval of time dictated by the Uncertainty Principle, there is no net creation of matter. Because these particles cannot be seen directly, they are 'virtual' rather than 'real'. Even though they cannot be seen directly, virtual particles are expected to endow the otherwise empty vacuum with a non-zero energy – vacuum energy. Consequently, rather than thinking of the vacuum as being mere 'nothingness' (an absence of 'stuff'), physicists regard it as the lowest energy state of space.

But if dark energy is indeed vacuum energy, there are two aspects of it that perplex physicists. Why is the energy density associated with the cosmological constant so exceedingly low? And why has it begun to dominate the dynamics of the universe so recently in the history of the universe? The first question forms the basis of the fine-tuning problem, and the second is the coincidence problem.

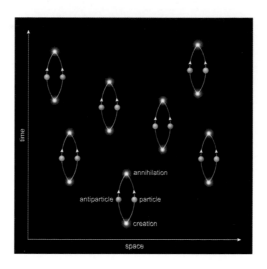

'Empty' space (the vacuum) is filled with virtual particles that exist so fleetingly that they cannot be seen directly.

Fine tuning and coincidence

When theoreticians try to calculate the expected density of vacuum energy, they obtain rather startling results. According to the simplest quantum physics calculations, the energy density of the vacuum should be about 10^{120} (1 followed by 120 zeroes) times greater than it actually is (assuming that vacuum energy contributes 70 percent of the critical density of the universe). If the vacuum energy density were really that high, the universe would blow up so fast that all forms of matter would instantly be ripped apart, and there would be no stars, galaxies or people. Because stars, galaxies and people do exist, some as yet undiscovered mechanism must cancel out all but a miniscule fraction of the vacuum energy – reducing it to just one part in 10^{120} of what it otherwise would be. To get a feel for just how big the discrepancy between theory and observation is – and hence the amount of 'cancellation' that is needed to match the two together – try writing down zero, followed by the decimal point, followed by 119 zeroes, then the numeral one (0.0 … 01); that represents the proportion of quantum vacuum energy left over after the putative cancellation process. Whereas theoreticians are usually sanguine about factor of two or even factor of ten (one order of magnitude) discrepancies between theory and observation, 120 orders of magnitude is rather extreme by anybody's standards!

The alternative is to suppose that the various factors that contribute to the vacuum energy cancel out completely, so that its net energy density, and hence the value of the cosmological constant, is precisely zero. Indeed, in some ways it may be easier to devise a theory in which the vacuum energy is exactly zero than to conjure up a process which cancels out all but one part in 10^{120} of what otherwise would be a truly colossal energy density. But if dark energy is indeed vacuum energy, then the initial properties of the universe had to be set up in such a way that the energy density of the vacuum was not only infinitesimally small compared to the initial densities of matter and radiation, but also had precisely the right value to ensure that it would become the dominant energy component some 8 billion years later. This precise fixing of its initial energy density (and the fantastic degree of cancellation needed to make it so feeble compared to what quantum calculations would suggest) constitutes the fine-tuning problem.

The other issue – the coincidence problem – is the focus of much collective angst among cosmologists. Why is the acceleration happening now – and why did it kick in so relatively recently? Or, to put the issue another way, why is the density of dark energy so closely similar to the density of matter and radiation at this particular time?

To see why this should be a problem, let's think about what happens to matter and radiation as the universe expands. The very early universe contained a densely-packed mix of photons and matter particles and its dynamics were controlled by the dominant influence of radiation (high-energy photons). At some particular instant, a representative region of space would contain a certain number of matter particles and photons. Because the volume of a three-dimensional object is proportional to the cube of its diameter, then each time the expansion of the universe stretched the diameter of that region of space by a factor of two, its volume would increase by a factor of eight ($2^3 = 8$) and the number of matter particles and photons in each cubic metre would

reduce by a factor of eight. But whereas the density of matter decreases in proportion to the cube of the expansion factor the energy density of radiation decreases with the fourth power of the expansion factor. The reason for this is that, not only does the number of photons in each cubic metre of space decline with the cube of the expansion factor, but radiation is also redshifted by the stretching of space, so that the energy associated with each photon is correspondingly reduced. Each time the size of the universe is doubled, the wavelength of a photon is doubled and its energy halved.

As the universe expanded and cooled down, the energy density of radiation declined faster than the density of matter so that, a few tens of thousands of years after the beginning of time, the density of radiation dropped below that of matter, and the universe shifted from a radiation-dominated phase to a matter-dominated phase during which the dominant energy density (the density of matter) continued to decline with the cube of the expansion factor.

At present the densities of dark energy and matter are remarkably similar. As we have seen, current measurements indicate that dark energy makes up slightly more than 70 percent of the total mass-energy density and matter somewhat less than 30 percent, so that the dark energy density is currently between two and three times as great as the matter density – a strikingly close match given the potentially enormous range of values that these two quantities could have possessed. If the dark energy component of the universe were vacuum energy (the cosmological constant), its energy density would have been exactly the same in the remote past (and will remain the same in the far future) as it is now. By contrast, the matter density, which is inversely proportional to the cube of the scale factor, would have been very much greater in the distant past. If the two densities are almost identical now, the ratio of vacuum energy density to matter density would have been approximately 1/1000 when the universe was one-tenth of its present size, 1/1,000,000 when the universe was one hundredth of its present scale, and 1/1,000,000,000 when the universe was about one thousandth of its present scale – round about the time when the microwave background radiation was released. At even earlier times – especially during the radiation era – the density of vacuum energy would have been utterly microscopic compared to the colossal densities of matter and radiation. Conversely, when the universe expands to 10 times its present scale, the unchanging density of vacuum energy will be 1000 times greater than the rapidly dwindling density of matter.

Since those early times when matter was dominant, it was inevitable that at some stage in the evolution of the cosmos, its density would drop below the feeble but unchanging density of vacuum energy, and that dark energy, therefore, would become the dominant cosmic constituent. But given the huge range of possible values which the cosmological constant could have had, why should it just happen to have become dominant so close to the present epoch in the history of the universe? If it had become dominant significantly earlier, it would have interfered with structure formation, and we would not be here to debate the issue. On the other hand, if its energy density had been such that it would not become dominant until significantly later in the history of the universe, then its present contribution to the overall density would be negligible, and we would not be aware of it. Why now? Do we live at a special time in the history of the universe?

Ever since Copernicus dethroned the Earth from its central position in the cosmos, and Hubble dethroned the Milky Way from its unique 'island universe' status as the totality of the stellar universe, cosmologists have been decidedly, if not obsessively, wary of anything which appears to ascribe a special position (in space or time) to our own situation. Perhaps it is pure coincidence that we happen to be around at the time, shortly after the end of the matter-dominated era, when repulsive dark energy has just gained the upper hand and the universe has embarked on its headlong accelerating expansion. Maybe it just happened to be that way. However, physicists find this explanation unsatisfactory, and would like to find good reasons why the cosmological constant is so weak and has become dominant only relatively recently.

To try to make sense of the fine tuning and coincidence problems, some cosmologists have resorted to the anthropic cosmological principle – the notion, in its least contentious form, that we exist because conditions are suitable for our existence (if they were not, we would not be here to observe the universe and deliberate on its nature). In a stronger form, the 'principle' implies that the universe somehow selects the right set of conditions to allow the existence of sentient observers. Our existence certainly seems to hinge on a remarkably favourable and finely tuned set of physical parameters. Had the force of gravity been a shade weaker, matter might never have coalesced into galaxies, stars or planets; had it been a shade stronger, the universe might have collapsed into a Big Crunch before sentient life could have evolved. Had the values of the fundamental constants (such as the charge on the electron and proton), or the strengths of the other forces of nature (weak and strong nuclear, electromagnetic), differed from their current values by a small fraction, the nuclear reactions that power the stars would not have occurred and the heavy elements needed to make planets – and us – would never have been forged. And, had the value of the cosmological constant, and the energy density of the vacuum, been substantially higher than it is, acceleration would have kicked in too soon, and the universe would have expanded far too quickly for galaxies, stars, planets or even atoms to form at all. Equally, we as a species could not have appeared on the scene until after the onset of structure formation and the birth and evolution of galaxies, stars and planets. The fact that we exist requires that the value of the cosmological constant be small, and that the densities of matter and dark energy be closely compara-

The multiverse hypothesis (left) suggests there may be a vast number of separate universes, whereas the 'bubble universe' concept suggests that there may be a multitude of different domains ('bubbles') within a single immense universe. In either case, our observable universe would be a tiny region within an individual universe or domain (A). Only in a very few universes or domains would conditions be right for sentient life to exist.

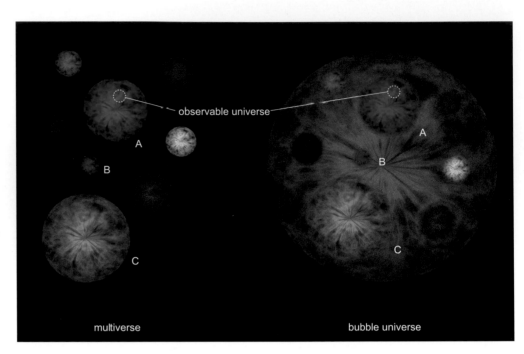

ble at the present time.

One possible anthropic explanation of why conditions in our universe are suitable for life, and specifically, why the value of the cosmological constant is as low as it is, is provided by the concept of the 'multiverse'. The essence of the idea – which has been discussed by, amongst others, Stanford cosmologist Andrei Linde and Cambridge cosmologist Martin Rees – is that there exists a vast (perhaps infinite) number of entirely separate and unconnected universes, each of which has different properties (for example, the constants of nature have different values, forces have different strengths and, in particular, the values of vacuum energy density are different). Only in a tiny proportion of these universes will conditions be suitable for the emergence and existence of sentient life, and ours happens to be one of that small band. After all, if our universe began as a tiny random quantum fluctuation that inflated rapidly, creating its own space and matter as it went, there is no reason why this event should have been unique. An alternative possibility is that inflation may have divided the universe into countless distinct domains with different properties – some of which might be suitable for life, but most of which would not – but

separated from each other by distances which are far greater than the observable horizon (the radius of the observable universe). If that is so, we live in one small part of one of this vast multitude of diverse bubbles, one that happens to have the right conditions for our existence and within which vacuum energy has a suitably low density.

Such a possibility is raised by string theory – an approach to unifying the forces of nature – which treats particles as different excitations of miniscule vibrating strings. Within string theory, there is potentially a truly enormous number of different possible states into which a system could settle, most of which would have a much higher value of vacuum energy than we see around us. If this approach is correct, the greater universe consists of a myriad of sub-universes each with a different value of vacuum energy. Regions with large positive values of vacuum energy would expand so violently that no structure or sentient beings could ever have formed. Those with negative values would rapidly collapse and vanish. Only those bubbles within which the vacuum energy is very close indeed to the observed value, could be suitable for life. We live in a region where the 'right' values pertain.

Although the multiverse and multiple domain concepts do indeed provide ways of producing a universe which, like ours, has the right conditions for life and the 'right' value for the vacuum energy, to many cosmologists these approaches seem decidedly unsatisfactory – a sort of cosmic 'cop out' which means that we may never be able calculate the precise observed properties of the universe directly from a complete physical theory. Perhaps this is the explanation, but one can't help feeling that perhaps nature could have found a better way to produce a universe where galaxies and sentient observers can exist than the 'scatter gun' approach of generating a vast horde of universes, or sub-universes, with random properties.

Evolving dark energy

Those who find the fine tuning and coincidence aspects of the cosmological constant/vacuum energy model highly troublesome have attempted to devise alternative formulations of dark energy which go some way towards explaining why the energy density of dark energy is as it is today. Surely, it has been argued, it would be more natural for dark energy to have started out in the early universe with an energy density similar in magnitude to the densities of matter and radiation and then to have evolved as the universe expanded. Rather than starting from an extraordinarily low value and remaining constant ever after (as for the cosmological constant/vacuum energy), evolving dark energy would decline at similar rates to matter and radiation densities, with dark energy only overtaking radiation and matter densities after structure had formed in the universe, and therefore only becoming dominant relatively recently.

One leading contender is quintessence, a type of all-pervading, repulsive dark energy that varies with time (and possibly from place to place), which was proposed in 1998 by Paul J Steinhardt, Rahul Dave and Robert R Caldwell, all of whom, at that time, were based at the University of Pennsylvania. The name 'quintessence' is rather appropriate,

for, historically, it alludes to the hypothesised 'fifth element' of Greek philosophy, the pure incorruptible substance of which the eternal and unchanging heavenly bodies were assumed to be composed. By contrast, the everyday world of human experience, where change and decay were rife, was composed of the elements earth, air, fire, and water. In its modern conception, quintessence is something very different: a type of quantum field (a scalar field – see 'Fields of various kinds' on page 146) which fills all space, has negative pressure and, crucially, varies with time.

Whereas such a field can take a variety of forms, a vital criterion is that its energy density should remain well below the energy densities of matter and radiation for much of the history of the universe and should become dominant only relatively recently. Of particular interest is a class of fields, called tracker fields, which arise in a number of theories that attempt to unify the forces of nature.

A key feature of tracker quintessence is that its energy density would 'track', that is, decline at the same rate as, the dominant energy component of the universe. As we have seen, the universe was dominated at first by energetic radiation but, because the energy density of radiation declines faster than the energy density of matter, it became dominated by matter a few tens of thousands of years after the Big Bang. Tracker quintessence would follow the decline of radiation density during the radiation-dominated phase, and the decline of matter density during the matter-dominated era. While the field tracks radiation or matter, its contribution to the overall energy density remains small and the expansion of the universe continues to slow down.

To initiate accelerating expansion, the field at some stage has to cease tracking the other densities and evolve towards a constant, or gently declining, energy value. When the matter density eventually drops below the quintessence density, the repulsive influence of quintessence drives the universe into acceleration. The problem is to arrange for the tracker field to become dominant at just

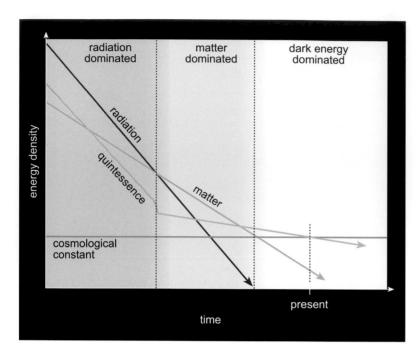

either by carefully choosing the mathematical form of quintessence and fine-tuning its parameters so that its density managed to overtake the matter density a few billion years ago, or by looking for some physical trigger that will change its rate of decline and ensure it becomes dominant at the 'right' time.

Perhaps the onset of dark energy dominance, and universal acceleration, was triggered by some natural event in the relatively recent history of the universe, such as the transition from the radiation era to the matter era and the onset of structure formation (matter could not begin to clump together until after the end of the radiation era). One promising possibility in this regard is k-essence (an abbreviation for 'kinetic energy-driven' quintessence), a particular form of tracker quintessence that was developed by C Armendariz-Picon and V Mukhanov of the Ludwig Maximilians Universität, in Munich, and Princeton cosmologist Paul J Steinhardt.

A distinctive property of k-essence is that tracking only occurs during the radiation-dominated era. When the densities of radiation and matter become equal, k-essence undergoes an abrupt transition from positive to negative pressure during which its energy density drops sharply by several orders of magnitude, and then begins to decline more slowly than matter density. K-essence cannot

The various lines show schematically how the densities of radiation, matter and two possible forms of dark energy change with time, leading to dark energy domination. Whereas the energy density of the cosmological constant remains constant, tracker quintessence eventually declines more gently than matter and radiation.

the right point to ensure that the universe embarked on its accelerating phase relatively recently. In that sense, quintessence still has to confront the coincidence problem. Although quintessence does not require fine-tuning of energy densities at early times, because it could evolve in many different ways from a wide range of initial energy densities, quintessence models still have to yield a ratio of dark energy to matter of about two to one in the universe today. This is achieved

Fields of various kinds

A field is, in a sense, a map of the way some particular quantity is distributed; for example, a magnetic field specifies at every individual point in space, the strength and direction of the magnetic force that would act on a charged particle (the direction in which the field acts at different points is charted out by 'field lines'). A field of this kind is a vector field (a vector is a quantity which has magnitude (i.e. 'size') and direction; for example, velocity is a vector). The type of field associated with quintessence (and with a broad range of alternative dark energy proposals) is a scalar field, where at every point the magnitude of a particular quantity is specified, but it has no directional properties (the term, scalar, refers to a quantity which has magnitude, i.e. 'size', but no particular direction). For example, atmospheric pressure and temperature are both scalars. Isobars on a weather chart, map out the 'pressure field' by joining up points at which the pressure is identical. Or, again, isotherms map out the 'temperature field' by joining up points where the temperature is the same. Whereas the pressure or temperature at a particular point does not have a direction, the pressure gradient (or temperature gradient) heads downward from high-pressure isobars (or high-temperature isotherms) towards low-pressure isobars (or isotherms). The gradients of these fields (in the absence of any other influences) determine the directions in which air (or heat) will flow.

dominate before the onset of the matter era because it is tracking the radiation density, nor can it dominate immediately after that time, because its energy density has dropped well below that of matter. However, because its energy density is then declining so much more slowly than that of matter, it must overtake matter not too long thereafter – at roughly the present time. This implies that the rise to dominance of the tracker field and the onset of structure formation are both triggered by the beginning of matter domination. In this scenario, it is not surprising to find that the universe should have begun to accelerate only in the relatively recent past after galaxies and stars have formed.

Quintessence models have equations of state in the range −1/3 to −1 (−1 < w < −1/3), and accelerate more gently than models in which the dark energy is a cosmological constant (for which w is constant and precisely equal to −1). As time goes by, and the density contribution of matter approaches ever closer to zero, then (always assuming that the sum of matter density and dark energy density is exactly equal to the critical density), the dark energy density becomes wholly dominant and – unless the nature of space and vacuum energy has further surprises in store – the acceleration will continue indefinitely thereafter. Because the rate at which the expansion of the universe accelerates depends on the precise ratio of the negative pressure of dark energy to its density (i.e. the equation of state, w), the way in which dark energy evolves (or doesn't, if it is a cosmological constant) influences the way in which the expansion rate of the universe changes with time. Cosmologists hope, therefore, eventually to be able to distinguish between different versions of dark energy, or at least to eliminate some of them, by measuring the expansion history of the universe with greater precision, and by putting tighter observational bounds on the dark energy equation of state (see Chapter 12).

The phantom menace

In an accelerating universe, galaxies will con-

tinue to fly away from each other at an ever-increasing rate, but the conventional view is that gravity will prevent galaxies themselves from being torn apart by the accelerating expansion of the universe. Systems that are currently bound together by gravity will continue so to be.

However, Robert R Caldwell, of Dartmouth College, New Hampshire, and Marc Kamionkowski and Nevin N Weinberg, of the California Institute of Technology, have come up with an altogether more disturbing possibility. In a paper published in 2003, they worked out what would happen if the dark energy that is driving cosmic acceleration were of a type that has been dubbed 'phantom energy'. Because the ratio of its negative pressure to its density is more strongly negative than −1 (i.e. w < −1), phantom energy behaves in thoroughly alarming fashion. Whereas the energy density of quintessence decreases as the universe expands, and the energy density of the cosmological constant remains the same, phantom energy, and the resulting acceleration, increases – slowly at first, but ever-faster – and becomes infinitely great in a finite time. In a universe dominated by phantom energy, the stretching of space eventually will become so severe that structures which are currently held together

The various curves show how the rate of cosmic expansion would change with time if dark energy were the cosmological constant (w = -1), quintessence (w = - 0.8 is a representative example), or phantom energy (w = - 1.2 is a representative example). Evolution with no dark energy is shown for comparison.

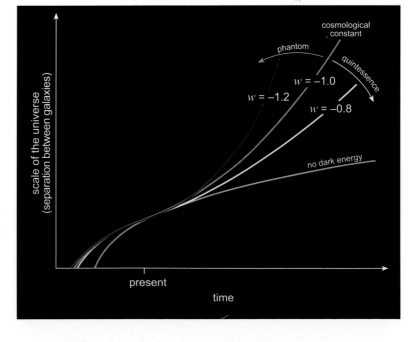

by gravity, the electromagnetic force or the strong nuclear force will be torn apart. Finally, the fabric of space itself will be rent asunder in a catastrophic 'Big Rip'.

The time interval from now to the Big Rip would depend on the precise value of the ratio of negative pressure to density. For one representative value of this ratio (−3/2), Caldwell, Kamionkowski and Weinberg have calculated that the Big Rip would take place in 22 billion years' time. In that particular case, the countdown to the Big Rip would run as follows. Galaxy clusters would be erased about a billion years before the Big Rip. Sixty million years before the final catastrophe, the Milky Way Galaxy would be pulled apart. Three months before the end

of time, the Earth would be torn from its orbit around the Sun, and 30 minutes before the end, the Earth itself would disintegrate. Molecules and atoms would be torn apart about 0.0000000000000000001 seconds (10^{-19} seconds) before the Big Rip, and in the brief instant that remained, atomic nuclei and nucleons would themselves be shredded. A more detailed calculation by S Nesseris and Leandros Perivolaropoulos, of the University of Ioannina, Greece, which was published in 2004, indicates that these disruptions would happen somewhat earlier. For example, for $w = -3/2$, the tearing apart of the Milky Way would occur about 180 million years before the Big Rip rather than 60 million years before (just when you thought the news

If dominated by the cosmological constant or quintessence, the universe will experience ongoing accelerating expansion. Phantom energy leads to a catastrophic Big Rip whereas if dark energy dwindles and changes sign, a Big Crunch lies in store.

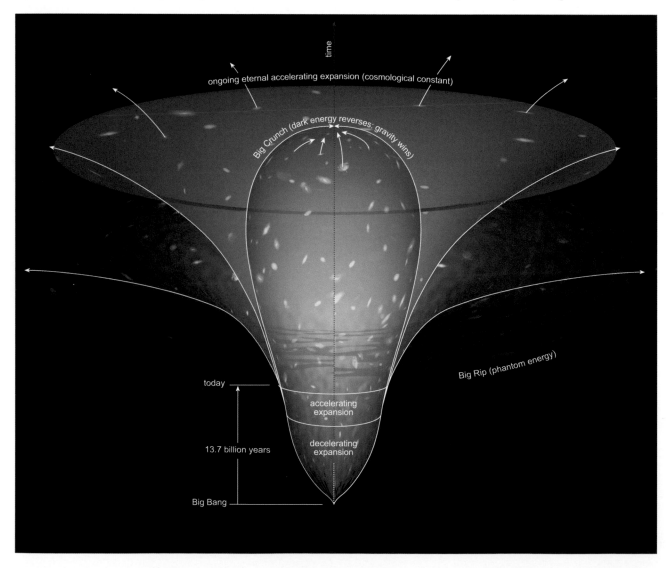

couldn't get any worse!).

The dominance of phantom energy would bring about the end of the brief era of cosmic structure, thus rendering the coincidence problem somewhat irrelevant, in the sense that the only time that structures, and ourselves, could exist, is close to the onset of acceleration. But in the face of this dire prediction, it is comforting to note that Madrid-based theoretician, Pedro Gonzales-Diaz, has shown that certain forms of phantom energy (based on Chaplygin gas models, of which more later) behave in a less destructive fashion and do not necessarily lead to a catastrophic Big Rip.

Big Crunch revisited

Since theoreticians as yet have no clear idea of the nature and properties of dark energy, they cannot rule out the possibility that dark energy might escalate out of control to cause a Big Rip. Nor can they exclude the possibility that dark energy might eventually just fade away, or even reverse its sign so as to become, like gravity, attractive rather than repulsive. In the former case, the acceleration would eventually cease, to be replaced – just as it was when the exceedingly brief era of inflationary expansion came to an end, a microscopic instant after the beginning of time – by gently-decelerating expansion under the control of gravity. But if dark energy were to change sign, a fate completely opposite to, but just as calamitous as, the Big Rip would lie in store. With gravity and dark energy pulling together, the universe would cease to expand, and would collapse rather quickly thereafter. Galaxies (or their remnants) would pile in on each other, and the entire universe would plunge into an all-embracing Big Crunch that would spell the end of space, time and matter.

In their attempts to get a handle on dark energy, theoreticians have explored many different models, some of which would drive the universe to this terminal catastrophe. For example, California-based cosmologists, Renata Kallosh, Jan Kratochvil and Andrei Linde (Stanford University), Eric V Linder

(Lawrence Berkeley National Laboratory) and Marina Shmakova (Stanford Linear Accelerator Center) have investigated a simple model (a 'linear' model) in which dark energy steadily declines so that it eventually becomes negative. In a paper published in 2003, they showed that in those circumstances even a flat universe could collapse into a Big Crunch within the next few tens of billions of years, though mercifully – on the basis of observational data available to the authors at that time – no sooner than 11 billion years hence!

Together with Yun Wang of the University of Oklahoma, Kratochvil, Linde and Shmakova took the idea a stage further in the autumn of 2004. Using the most recent supernova data, together with cosmic background measurements from WMAP and ground-based instruments, along with large-scale structure data from the 2dFGRS, they were able to put firmer constraints on the time to cosmic doomsday (the Big Crunch). Their conclusion was that, for this particular model of dark energy, our universe probably wouldn't collapse earlier than 24 billion years hence, but it will most probably live no longer than 3650 billion years. More cosmological observations and a deeper understanding of the theory of dark energy will be needed in order to improve constraints on cosmic doomsday time, if indeed the Big Crunch is the fate that lies in store.

The coincidence problem again!

Curiously enough, it appears that the coincidence problem could be greatly ameliorated in Big Rip or Big Crunch models in which the total lifetime of the universe is not much longer than the age of the universe today. In the former case, the expansion rate slows down as dilution of matter weakens its gravitational influence, and then accelerates catastrophically in its later stages when the influence of repulsive dark energy becomes overwhelming. In the latter scenario, the accelerating phase is brought to an end when the influence of dark energy becomes attractive rather than repulsive, and rapid collapse ensues. Robert J Scherrer of Vanderbilt

University, Tennessee has pointed out that because the Big Rip is triggered by the onset of phantom dark energy domination, the universe must spend a significant fraction of its total lifetime in a state in which the dark energy and matter densities are roughly comparable. Whereas the two densities differ from each other by a factor of about two at the moment, he argues that even a factor of 10 would certainly be regarded as 'coinciden-tal', and in a paper published in 2004 goes on to show that for values of w between –1.5 and –1.1, the two densities remain within a factor of 10 of each other for between one-third and one-eighth of the total lifetime of the universe.

In complementary vein, Pedro P Avelino, of the Universidade do Porto, Portugal, has investigated the coincidence issue within the context of universe models where the acceler-ation phase is followed by rapid collapse into a Big Crunch. By making some reasonable assumptions he found that our existence at this particular time – a few billion years after the onset of acceleration – is not a particularly strong 'coincidence' provided that the present value of the dark energy equation of state (w) is greater than (i.e. less negative than) –0.98, and the total lifetime of the universe is less than about ten times its present age.

Further possible solutions to the coinci-dence, or 'special time' problem, are offered by models in which periods of matter domi-nation alternate with periods of dark energy domination, either because dark energy is some kind of oscillatory field, or because periods of dark energy domination arise from different quintessence fields which dominate at different times. In the context of the former possibility, Scott Dodelson, Manoj Kaplinghat and Ewan Stewart, of Fermilab and the University of Chicago, back in the year 2000, posed the question: what if the universe has been accelerating periodically in the past? If so, it would not be particularly surprising that the universe happens to be accelerating just now. If the dark energy tracks, but oscillates above and below, the ambient background energy density (of radiation, matter, or what-

ever) then the universe would experience alternating bouts of acceleration and decel-eration. Some fine-tuning is still required, though, to ensure that the model would end up with the right properties today.

As an example of the latter, Kim Griest of the University of California, San Diego, has proposed that it may be possible to remove the cosmic coincidence problem by posit-ing not just one scalar field whose vacuum energy has become dominant recently, but a whole ensemble of fields whose energies span a huge range, some of which may have dominated the universe for short periods of time at different times in the past. If that were so, the specific type of dark energy that is responsible for the current acceleration would not be special. The field that was responsible for inflation (which is commonly referred to as the 'inflation field') in the early universe may be one of this ensemble; the Higgs field (which in the standard particle model gives mass to the W and Z bosons) would be another.

If so, the current phase of acceleration is temporary and will eventually come to an end. There may have been several periods of vacuum-energy dominated acceleration in the past, each followed by a period of radiation or matter domination, and there may well be further bouts of accelerated expansion in the future. If this proposal is correct, the ultimate fate of the universe cannot be predicted with-out understanding the origin and energies of all of these fields.

What happens at the end of each bout of vacuum energy domination? The current bout of acceleration, driven by an as yet unknown form of dark energy, has arisen after a long period of matter domination which itself was preceded by the dramatic, though short-lived, accelerating inflationary era. At the end of inflation energy converted to particles of matter and antimatter, and photons. Before inflation, the high-energy vacuum – known as 'false vacuum' – was not the lowest possible energy state. Inflation occurred during the transition from the high-energy 'false' vacuum to a lower-energy vacuum state. But what if the present vacuum is not the lowest possible energy state? The current phase of acceleration may be a symptom of a slower and longer-lived phase transition that will come to an end when the vacuum has dropped to an even lower stable state. The end of that phase may be marked by the creation of new forms of particles, matter and structure. Further bouts of matter domination and dark energy domination – on successively feebler energy scales – may follow.

A proliferation of theories

The discovery of dark energy and cosmic acceleration has turned our understanding of the universe on its head, but our ignorance of its nature remains profound. Not surprisingly, therefore, the minds of cosmologists, particle physicists and theoreticians have gone firmly into overdrive in their efforts to find a viable theory of this perplexing new cosmic ingredient. Although the cosmological constant (vacuum energy) and quintessence are the current front-runners, and phantom dark energy is also very much in the frame, theoreticians have already come up with many other possibilities.

One of the more intriguing suggestions is that if the universe has more than four dimensions (as string theory, for example, implies), then gravity may leak away from our universe into higher dimensions, so weakening its effect. This would explain why gravity is so much feebler than the other forces of nature. Furthermore, as the inherent strength of gravity weakens, cosmic expansion speeds up, giving the impression that a repulsive force is driving it. Others have looked at the possibility that the cosmological constant, Λ, is not a true 'constant' at all, but may instead change over time. For example, Joan Solà and Hrvoje Štefančić, of the University of Barcelona, have recently proposed a model in which the vacuum energy density of the cosmological constant declines, but the ratio of vacuum energy density to pressure remains the same ($w = -1$) so that it still behaves like a true cosmological constant at any particular time in the history of the universe. Their model

(which could also involve a variation in the gravitational constant, G, which determines the strength of gravity) appears to be able to mimic the effects of quintessence and phantom energy. Among others who have looked at the possibility that the strength of gravity may change with time (an idea that was first mooted by Cambridge physicist Paul Dirac in 1937), Indian physicists Saibal Ray and Utpal Mukhopadhyay have suggested that if Λ is declining then, depending on whether G decreases or increases over time, the ultimate outcome could be eternal accelerating expansion or an eventual Big Crunch.

Another idea that is attracting a lot of interest is the notion that dark energy may have started out behaving like quintessence (with w less negative than –1) then evolved across the so-called 'cosmological constant barrier' and begun to behave like phantom energy (with w more negative than –1) relatively recently. Chinese theoreticians Bo Feng, Xiulian Wang and Xinmin Zhang have suggested that this could happen if dark energy consists of two components – quintessence and phantom energy – the combination of which they have called 'Quintom'. Feng, Zhang and their co-workers have developed an oscillating Quintom model in which the energy density oscillates over long periods of time, this leading to the ongoing expansion of the universe with periodic episodes of phantom-dominated acceleration – the early inflationary epoch and the current bout of acceleration being two such episodes. Like some of the other oscillating dark energy models described earlier, oscillating Quintom would alleviate the vexed 'coincidence problem' because it implies that, instead of living peculiarly close to the beginning of a unique period of acceleration driven by the onset of dark energy domination, we just happen to be around during one of a whole series of episodes of this kind. Oscillating Quintom would not lead to a Big Rip or to a Big Crunch.

It is also conceivable that dark energy could take the form of relic particles left over from the Big Bang, provided that any such particles are light enough still to be moving around so fast (i.e. to be sufficiently 'relativistic') that gravity cannot pull them together to form localised clumps. This possibility was recognised in 1997 by Michael S Turner and Martin White of the University of Chicago. On the basis of the information available at that time about the mean matter density and the sizes of the most prominent temperature fluctuations in the microwave background, and on the theoretical merits of inflationary theory and cold dark matter, Turner and White had argued the need for an additional 'smooth component' in the mass-energy density of the universe to augment cold dark matter and bring the density up to critical. As an alternative to the cosmological constant, they suggested various possibilities, one of which was relativistic particles, which they labelled the 'X-component', or 'X-matter'.

Recently, a number of theoreticians have turned their attention to models in which dark energy is linked to neutrinos. This interest has been stimulated by the curious fact (another 'coincidence'?) that the level of mass-energy associated with neutrinos is similar to that of dark energy. In particular, Rob Fardon, Ann E Nelson and Neal Weiner, of the University of Washington, have proposed that dark energy may be linked to neutrinos whose masses vary (mass-varying neutrinos, or MaVaNs). They argue that MaVaNs can behave as a negative pressure 'fluid' which could be the origin of cosmic acceleration. They suggest that the masses of neutrinos depend on the number density of neutrinos (the number of neutrinos in each cubic metre of space), and that neutrino masses increase as the universe expands and becomes more dilute.

Their proposal is that neutrinos interact through a new scalar field called the 'acceleron' which tracks the density of neutrinos and ensures that this form of dark energy density remains similar to the energy density of neutrinos for a large part of cosmic history. According to the theory, this new 'dark force', felt by neutrinos, is what drives cosmic acceleration. Because neutrino masses would have been much lower in the past, their contribution to the total density of the early universe would have been much smaller than the contributions due to baryons, radiation or cold dark matter. However, as the ongoing expansion of the universe rapidly diluted the gravitational influence of these other components, the influence of varying mass neutrinos and the acceleron field became progressively more important.

The theory makes some potentially testable predictions, in particular: MaVaN masses should be no greater than about 10^{-4} eV/c^2 (to prevent them from clumping); in addition to the three known neutrino types, there should be extra 'sterile' neutrinos (neutrinos which only take part in flavour mixing – see Chapter 5); and, if cosmic evolution and environment influence neutrino masses, then astrophysical and cosmological measurements of neutrino masses may differ from those obtained in terrestrial experiments. So, perhaps the humble neutrino, which in recent years has been relegated to being little more than a minor player in the grand scheme of cosmic evolution, may after all turn out to have a decisive role to play as a result of its interplay with the hypothesised acceleron field.

Two sides of the same coin?

Various research groups have been exploring the possibility of unified dark energy (or 'quartessence') models in which dark matter and dark energy are seen to be different aspects of the same thing. Unified dark energy models are often discussed in the context of a 'Chaplygin gas', a type of 'fluid' (which is assumed to fill all space) which has the distinctive property that the lower the density, the greater the negative pressure. The concept was devised, in a completely different context, by Russian hydrodynamicist Sergei Chaplygin, about a century ago. Because of this property, the fluid can mimic the behaviour of matter very early in the history of the universe. When density is high, the pressure becomes negligible, so that at early times the equation of state becomes essentially $w = 0$, which is the equation of state of cold dark matter. Later in its history, when the density is low, the pressure becomes strongly negative (which is the essential property of repulsive dark energy).

Others have explored ways of coupling matter and quintessence together so that energy can be transferred between them. For example, Greg Huey and Benjamin D Wandelt, of the University of Illinois, have devised an interacting quintessence cold dark matter (iQCDM) model, based on string theory, which posits the existence of several different components of dark matter, only one of which (iCDM) interacts with quintessence. As baryons and non-interacting dark matter components become ever more dilutely spread out due to the expansion of the universe, the transfer of energy from quintessence to interacting dark matter ensures that iCDM becomes significant around the onset of quintessence domination, and ensures that the densities of interacting cold dark matter and quintessence evolve towards a constant ratio. They suggest that this type of model can protect the universe from premature acceleration (so allowing structures to form), evolve to give a constant fraction of dark energy to dark matter (thereby solving the coincidence problem), and avoid a Big Rip.

Or, again, Salvadore Capozziello (of the Università di Napoli, Italy) and colleagues have tried to incorporate dark matter into a unified cosmology in which the same evolving scalar field, at different times in the history of the universe, corresponds to the inflation field that drove inflation in the very early universe, quintessence at intermediate times, and phantom dark energy at late times. And, on a different tack, University of Michigan theoretical physicists J G Hao and R Akhoury, have begun to explore the possibility that Bekenstein's relativistic MOND theory (see Chapter 6), which involves extra fields in addition to gravitation, may be able to reproduce the effects that are usually attributed to dark matter and dark energy without directly invoking either of them.

If any of these ideas are heading in the right direction, we may yet find that dark matter, which clumps together in galaxies, clus-

ters and other forms of large-scale structure, may be intimately linked to – and may simply be another face of – the apparently smoothly-distributed component which we call dark energy. To account for 95 percent of the content of the universe we may in the end have to deal with just one (admittedly, more complex) mystery ingredient instead of two.

Where do we stand on dark energy today?

The existence of dark energy seems to be firmly established by a wide range of observational evidence, but cosmologists and physicists remain very much 'in the dark' about its physical nature. The cosmological constant (in the guise of constant vacuum energy), and various forms of quintessence (dark energy which varies with time) are the leading contenders, but the richly varied plethora of alternative possibilities – of which there are many more than have been described in this chapter – testifies to the significance of the quest to understand dark energy, and to the fertile imaginations and inventiveness of theoreticians.

When cosmologists wakened to the possibility that dark energy might vary with time, or from place to place, or might even interact or interchange in some way with dark matter, they opened up a Pandora's box of possibilities, ranging from smooth perpetual accelerating expansion, to alternating bouts of acceleration and deceleration, through to the opposing extremes of the Big Rip and the Big Crunch. All of these possibilities will remain in the frame until such time as cosmologists, or particle physicists, can pin down the properties of dark energy, and the way in which it evolves, with sufficient precision to eliminate at least some of them.

Testing the New Cosmology

The current Standard Model against which others are judged, is ΛCDM – a critical-density, geometrically flat universe filled with vacuum energy (the cosmological constant, Λ) and cold dark matter. According to this model, the seeds from which galaxies and large-scale structures grew were microscopic quantum fluctuations that were stretched to cosmic dimensions by a brief but spectacular bout of inflationary expansion that took place a miniscule fraction of a second after the beginning of time. The warmer and cooler patches that we observe in the cosmic microwave background are the visible imprint of density fluctuations – regions within the primordial mix of matter and radiation which were denser, and less dense, than average. After the decoupling of matter and radiation, which took place some 380,000 years after the Big Bang, these fluctuations evolved under the influence of gravity to form the rich variety of structures that we see in the universe today. Galaxy formation proceeded from the bottom up, with cold dark matter playing a vital role. Smaller clumps in the underlying distribution of dark matter collapsed first, forming the gravitational 'wells' into which baryonic matter subsequently fell to create the protogalaxies which, through a succession of mergers, coalesced into successively larger galaxies and clusters. It's an impressively comprehensive picture.

ΛCDM has proved to be a very successful model, but it is there to be tested and challenged. Despite its great achievements, it has problems and issues to contend with. For example, deep space surveys have revealed galaxies and clusters in the very early universe which appear to have formed sooner and evolved faster than the Standard Model would predict. In particular, some of the very high-redshift galaxies revealed by the Spitzer orbiting infrared observatory, which are seen as they were when the universe was less than 1 billion years old, appear to contain populations of cool red stars that were not expected to have evolved to that stage in their lives so early in the history of the universe.

Furthermore, the microwave background data obtained from WMAP's first year all-sky survey appears to contain anomalies – anomalies which seem still to be present in the more precise 3-year temperature maps that were released in March 2006 - which some cosmologists reckon could be a ticking time-bomb for the standard inflationary model. According to that model, the distribution of warmer and cooler patches, which delineate the underlying density fluctuations, should be random and look the same in all directions across the whole sky. Yet, several different groups of cosmologists have reported finding indications of a non-random alignment in the patterns of spots, and an apparently significant north-south difference in the detailed appearance of the microwave sky on opposite hemispheres of the sky. This is a difference which may be linked to the presence of what seems to be a particularly large and exceptionally cool spot that lies well to the south of the galactic plane.

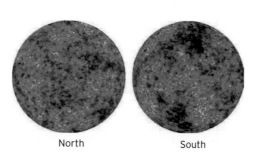

North   South

Based on the first three years of WMAP data, these maps show the distribution of hot (red) and cold (blue) spots in opposite hemispheres of the sky: north of the plane of the Earth's orbit (left) and south of the plane of the Earth's orbit (right). There appear to be significant and puzzling differences between the two hemispheres.

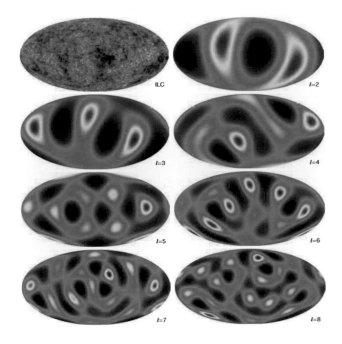

ILC   l=2
l=3   l=4
l=5   l=6
l=7   l=8

The microwave sky contains some puzzling features. Computed from the WMAP all-sky temperature map (labelled 'ILC', top left) the other maps display large-scale temperature features ('modes') of different angular size (labelled from 'l = 2' to 'l = 8' in order of decreasing angular size) which are present in the data. Some of these maps display peculiar patterns. For example, the l = 2 and l = 3 modes appear to be strongly aligned whereas the l = 5 mode appears remarkably symmetrical and non-random.

The claimed alignments show up in the distribution of some the largest structures in the microwave all-sky maps (features which – because of their very great sizes – are believed to have changed relatively little since the inflationary era and which, therefore, ought to represent the 'cleanest' fossilised imprints of the primordial density fluctuations that were stretched by a colossal factor a microscopic fraction of a second after the Big Bang. These features appear to display a degree of common alignment along an axis (which has been dubbed 'the Axis of Evil') which points in the general direction of the Virgo cluster of galaxies and roughly at right angles to the plane of the Earth's orbit round the Sun. These curious anomalies have led some cosmologists to argue that the large-scale features on the temperature maps in part (or even wholly) may have been generated by 'local' processes related, perhaps, to nearby clusters of galaxies or even to the motion and environment of the Solar System. At the very least, the observations imply that the large-scale temperature maps are to some extent contaminated by 'foreground' effects, which makes their interpretation more difficult. At worst, the large-scale features may not be genuine primordial ones at all.

It is an immensely difficult task to identify and remove all of the different kinds of foreground contamination that are, or could be, present in the raw data – for example, radiation emitted by heated dust or electrons near to the plane of our own Galaxy, the scattering of background radiation photons as they pass through clouds of ionised gas in galaxy clusters (the S-Z effect), distortions produced by gravitational lensing, and effects produced by the shells of hot ionised gas that were created when the first stars switched on, a few hundred million years after the beginning of time. A plausible explanation is that the observed anomalies are artefacts which arise because experimenters have not yet succeeded in subtracting out all the unwanted foreground effects. Another possibility is they could be due in part to hitherto unidentified defects in the spacecraft's instrumentation, although in view of all the analyses and cross-checks that have taken place since WMAP began to return data, that seems exceedingly unlikely. Another intriguing possibility, suggested by Chris Vale of Fermilab and the University of California, is that the "Axis of Evil" effect is caused by the influence of weak gravitational lensing (distortions caused by the distribution of matter along different lines of sight) on the microwave background 'dipole' (as we saw in Chapter 8, the motion of the Solar System relative to the general background of microwave photons causes the sky in the direction in which the Solar System is heading to appear hotter than the opposite side of the sky; this phenomenon is called the CMB dipole).

But if they *are* genuine features in the primordial microwave background, their existence could force cosmologists to question the whole inflationary scenario and to rethink the origin and evolution of structure in the universe. Furthermore, if the background radiation is really telling us that the universe is *not* the same in every direction and that it contains the imprint of some special preferred direction, such a discovery would demolish one of the twin pillars of the cosmological principle on which conventional models of the universe are based – the principle which states that the universe on the large scale is homogeneous (is the same everywhere) and isotropic (looks the same in every direction). That would have wide-ranging ramifications for the whole of cosmology.

The resolution of that particular issue will require long and intensive analysis of existing and forthcoming microwave background data, from WMAP and from other sources – in particular the forthcoming Planck mission (of which more later). But for the moment, despite the odd crack here and there, 'standard' Big Bang cosmology, which incorporates inflation, cold dark matter and dark energy, is an impressively comprehensive edifice that appears to be holding together rather well.

To test it further, cosmologists are focusing on a number of key tasks: measuring the various cosmological parameters with better precision, measuring the expansion history of the universe, and establishing the precise sequence of events that transformed primordial density fluctuations, stage by stage, into the rich pattern of structure that we see in the universe today. They and particle physicists alike are desperate to identify what kinds of particles make up the inventory of non-baryonic dark matter and, above all, to find out the physical nature of the most perplexing enigma of all – dark energy. The nature of dark energy is arguably the deepest puzzle in present-day physics, the solving of which would revolutionise not only cosmology, but also the whole of physics.

Homing in on dark energy: 'w' holds the key

Is dark energy Einstein's cosmological constant in the guise of vacuum energy, or is it something more exotic such as quintessence or phantom energy? The key to sorting out which kind of dark energy is 'out there' is the ratio between its pressure and its density (the equation of state, w). As we saw in the previous chapter, if it is the cosmological constant, $w = -1$ and its value does not change with time. If it is 'something else' the value of w will be greater than (less negative than), or less than (more negative than) -1, and it may also change with time. The more negative the value

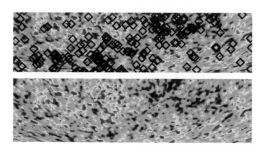

The upper panel shows part of the WMAP temperature map with diamonds representing the positions of nearby galaxy clusters. The lower panel shows the same area without the cluster positions. A University of Durham team has pointed out that clusters and superclusters tend to lie in cooler (blue) temperature regions, and that hot gas within these entities may have distorted the appearance of the microwave background to some extent.

of w, the stronger the negative pressure, and the greater the acceleration that dark energy will impart to the universe; the less negative the value of w, the gentler the acceleration. Furthermore, the more negative the value of w, the more rapidly its contribution to the overall density of the universe will decrease as we look out to greater and greater redshifts (and hence back in time towards the early universe), and that will reduce its influence on structure formation at early times. Because different types of dark energy affect the rate of cosmic expansion *and* the growth of structure in different ways, cosmologists hope to be able to pin down its properties by making more precise and detailed observations of the expansion history of the universe and the way in which structure has evolved over time.

One cornerstone of this endeavour will be deeper, wider, and more sensitive surveys of the distribution of luminous and dark matter. Looking outwards from Earth, we can divide space into concentric shells, each of which corresponds to a narrow range of redshifts and a different stage in the history of the universe (in practice, each shell has to be quite thick in order to pick up the largest structures). By mapping the distribution and degree of the clustering of matter in these various shells, cosmologists hope to build up a detailed history of the evolution of structure in the cosmos. The results of these surveys can then be compared with theoretical models and with huge numerical simulations, carried out in supercomputers, which calculate the gravitational interactions between vast numbers of point masses.

The largest and most elaborate simulation so far is the 'Millennium Simulation', which was carried out by the Virgo Consortium (a collaborative venture involving research teams from the UK, Germany, Canada and the USA) and published in the journal *Nature* in June 2005[1]. Assuming ΛCDM cosmology, this monumental computation followed the behaviour, under the influence of gravity, of about 10 billion point masses, in a cubic volume of space which (today) measures 2230 million light-years on each side, from a redshift of 127 (corresponding to when the scale of the universe, and the diameter of the simulation's sample volume, was 128 times smaller than it is today) up to the present day. According to its authors, this simulation is so powerful that it has succeeded in predicting the present-day locations, velocities and intrinsic properties of every single galaxy and dark matter halo more massive than the Small Magellanic Cloud (the smaller of our Galaxy's two companions) that lies within that volume of space. This is an immensely detailed theoretical reference against which to compare the observations.

Surveying the cosmos

Looking far out into space, the most distant 'shell' that we can study is the cosmic microwave background, the 'surface of last scattering' which reveals the state of matter and radiation at the time of decoupling, and which bears the fossilised imprint of the primordial density fluctuations as they were some 380,000 years after the beginning of time. That gives us an early starting point for mapping the growth of structure. Surveys of the distribution of galaxies in our own locality reveal the stage to which structures have evolved now (13.7 billion years after the beginning of time). By surveying more and more distant shells at progressively higher redshifts, cosmologists are hoping to eventually build a bridge that links the present-day universe to its distant past.

The most comprehensive galaxy distribution surveys so far are the Anglo-Australian Two degree Field Galaxy Redshift Survey (2dFGRS) which measured brightness, position and redshift data for more than 220,000 galaxies, and the ongoing Sloan Digital Sky Survey (SDSS), which is currently approaching its initial target of acquiring similar types of data for more than a million galaxies and around a hundred

These four frames from the Virgo Consortium's 'Millennium Simulation', running clockwise from top left, show stages in the evolution of dark matter structures in a region of space some 425 million light-years wide (assuming that the value of the Hubble constant is 70 km/s/Mpc), from about 200 million years after the Big Bang to the present time. Noted beside each frame is the time that has elapsed since the Big Bang expressed in billions of years (Gyr) and the cosmological redshift (*z*) corresponding to each frame.

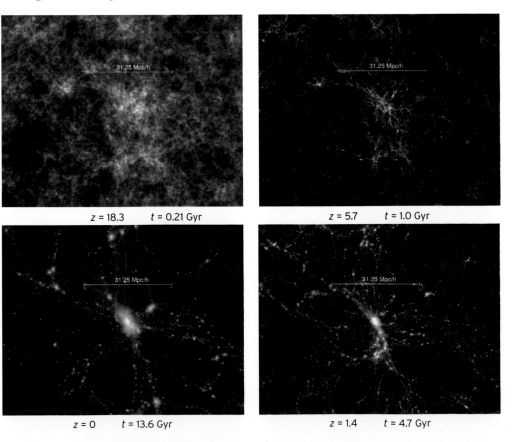

$z = 18.3$ $t = 0.21$ Gyr

$z = 5.7$ $t = 1.0$ Gyr

$z = 0$ $t = 13.6$ Gyr

$z = 1.4$ $t = 4.7$ Gyr

thousand quasars, and which is scheduled to continue (as SDSS II) until 2008. Both projects have produced detailed three-dimensional maps of the distribution of ordinary galaxies out to redshifts of 0.2–0.3 (which corresponds to distances of 3 billion light-years or more). In addition, the SDSS has mapped the distribution of a class of particularly luminous galaxies (called Luminous Red Galaxies) to a redshift of about 0.5. SDSS has also been able to detect quasars (which are exceedingly luminous) at much greater distances – out to redshifts in the region of 6.5 (which corresponds to seeing them as they were less than a billion years after the Big Bang).

Quasar light provides another means of probing the distribution of matter across great swathes of space and time. While en route towards the Earth, beams of light from these remote cosmic searchlights encounter numerous clouds of hydrogen gas, strung out along the line of sight. Atoms of hydrogen absorb ultraviolet light at a number of different wavelengths, most strongly at 121.6 nanometres, which is the wavelength of the Lyman-alpha

line in the absorption spectrum of hydrogen (see Chapter 1). Each cloud imprints its own Lyman-alpha line on the spectrum of the quasar. If a hydrogen cloud is stationary relative to the Earth, it absorbs light at precisely 121.6 nanometres and imprints a dark line on the quasar's spectrum at that wavelength, but if the cloud is receding, the imprinted line is redshifted to a longer wavelength. The hydrogen absorption in a quasar's spectrum is produced by a great many clouds (possibly hundreds) at different distances, each of which is receding from us (because of the expansion of the universe) at a different rate. Consequently, each cloud's absorption line is redshifted by a different amount, with the result that a great many absorption lines are imprinted at different wavelengths on the quasar's spectrum. The array of lines is called the 'Lyman-alpha forest'.

By measuring the amount of light absorbed (at the redshifted Lyman-alpha wavelength) by each of the clouds, astronomers can map the density, distribution and 'clumpiness' of gas at different redshifts along the line of sight towards each quasar. By studying large

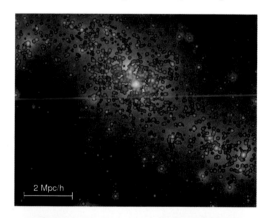

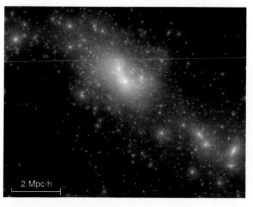

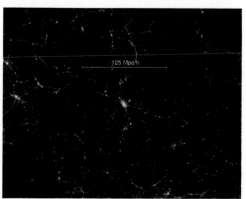

The top two images show the galaxy distribution in the Virgo Consortium's Millennium Simulation on very large scales (right) and on the scale of a rich cluster of galaxies (left), where individual galaxies can be seen. The top right panel, therefore, represents the large-scale distribution of light in the universe. The bottom two images show the corresponding dark matter distributions.

The Lyman-alpha forest arises because each cloud of gas along the line of sight absorbs ultraviolet light from a distant quasar. The observed wavelength of the 'Lyman-alpha' absorption lines that are imprinted on the quasar's spectrum depends on the redshift of each cloud. The quasar itself has a Lyman-alpha emission line that has been shifted from the ultraviolet to the extreme red end of the spectrum.

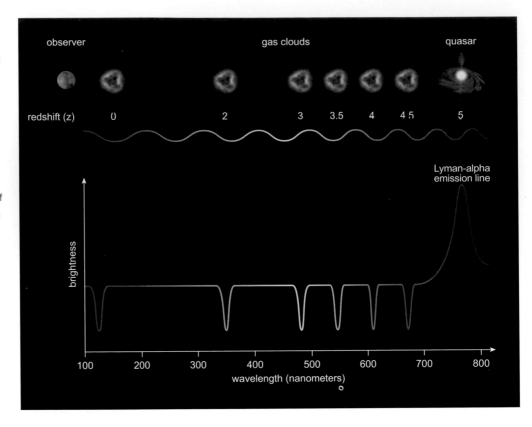

This tracing shows part of the Lyman-alpha forest (the large number of absorption lines which appear as dips in the plot of 'relative intensity') in the spectrum of the southern-hemisphere quasar, QSO HE2217-2818, which has a redshift of 2.4.

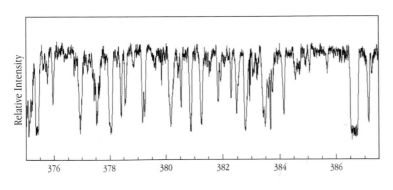

numbers of quasars, they can build up a comprehensive picture of the degree of clumping in the general distribution of hydrogen gas over a wide range of redshifts and look-back times into the history of the universe. Because quasars can be studied at much greater distances than ordinary galaxies, the Lyman-alpha data probes the distribution of clumps and filaments of gas at much higher redshifts (typically in the region of $z = 2$ to $z = 4$) than have been explored by galaxy redshift surveys, so allowing cosmologists to probe the evolution of structure much earlier in the history of the universe.

The real power of these various surveys emerges when their results are combined. Whereas galaxy distributions pick out structures ranging from tens of millions to many hundreds of millions of light-years in diameter, the hydrogen cloud observations reveal structures on smaller scales, from a few million to around a hundred million light-years across. The temperature maps of the microwave background reveal very large-scale density fluctuations which correspond (in the present-day universe) to structures that span hundreds of millions or billions of light-years. Between them, the Lyman-alpha clouds, the galaxy surveys and the microwave background maps, reveal the degree of matter clumping on scales ranging from a few million to several billion light-years. This information can be plotted as a spectrum of clumping, known as the matter power spectrum, which can be compared with the predictions of models which have different values of baryon density, dark matter density, dark energy density, and so on.

Many different research teams have combined different data sets to great effect. The following is just one representative example. In 2004, by combining Lyman-alpha forest data from more than 3000 quasars with WMAP temperature maps, SDSS galaxy data, and the latest available supernova results, a Sloan Digital Sky Survey team was able to derive some of the best values yet for some of the key cosmological parameters. They calculated values for the density contributions of matter and dark energy of $\Omega_M = 0.28 \pm 0.02$ and $\Omega_\Lambda = 0.72 \pm 0.02$, and showed that the observed pattern of mass clumping appears to match well with what inflationary theory predicts. Their deduced value for the dark energy equation of state (based on the evolution of matter clumping and the measured expansion history of the universe) was $w = -0.98 \pm 0.12$, with no evidence for any change with time, which is very much in line with the cosmological constant.

Furthermore, because neutrinos with finite masses would exert a detectable influence on the way in which structures developed in the early universe (the greater their masses, the more they would slow down the growth of small-scale structures), the combined observational data can be used to measure, or place upper limits on, the masses of neutrinos. From the Lyman-alpha, first-year WMAP and SDSS galaxy data, they found that the sum of the masses of the three known neutrino types (see Chapter 5) must be less than 0.42 eV/c².

Weak lensing has strong potential

Among the array of techniques for probing the large-scale distribution of mass in the universe, a comparatively 'new kid on the block' is weak gravitational lensing. As we saw in Chapter 3, rays of light travelling from remote background galaxies experience a succession of deflections, in assorted directions, as they pass by or through the various intervening concentrations of mass (both luminous and dark) that lie along the line of sight. The cumulative effect produces subtle statistical distortions (called 'cosmic shear') in their perceived shapes. Since the expected degree of distortion is only about 1 percent, it is hardly surprising that it was not until the year 2000 that any observing

groups succeeded in measuring the effect (see Chapter 3). The basic approach is to use remote faint blue galaxies – a population of galaxies which are seen as they were when the universe was about half its present age – as a kind of background 'wallpaper' against which to pick out patterns of distortion in their shapes and alignments. These patterns can then be converted to two-dimensional maps of the cumulative influence of all of the mass along the line of sight from the Earth to the background galaxies.

The trick is to measure the brightnesses of the background galaxies through a set of different colour filters and use that data to estimate their approximate redshifts (because redshift changes the wavelength at which a galaxy shines most brightly, a highly-redshifted galaxy will appear dimmer at short wavelengths and brighter at long wavelengths than a similar low-redshift galaxy). Redshifts obtained in this way are called 'photometric redshifts'. Astronomers can then identify all of the background galaxies that lie within a selected range of redshift, and measure the pattern of distortions for those galaxies alone. By repeating this exercise for a succession of shells, each spanning a different redshift range, they can build up a three-dimensional picture of how mass is distributed

In August 2006 Douglas Clowe's team in Arizona provided some of the strongest evidence that most matter in the universe is dark and non-baryonic.

This composite image from their work shows the galaxy cluster 1E 0657-558, which was formed when two large clusters of galaxies collided. The two pink clumps depict hot x-ray emitting gas, which contains most of the ordinary baryonic matter in the two clusters. The visible galaxies are shown in orange and white, and the blue areas show where most of the mass in the colliding galaxy clusters is concentrated.

If the colliding clusters consisted only of baryons, as MOND (in its basic form) requires, the observed separation of dark matter and x-ray emitting clouds would not have occurred.

 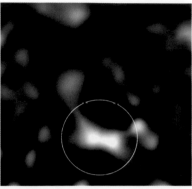

To the left is a deep image of a region of sky obtained by the ESO's Very Large Telescope. To the right is the reconstructed image of the distribution of mass (mainly dark matter) in this direction based on an analysis of gravitational weak lensing ('cosmic shear'); that is, the elongations and directions of the axes of galaxy images in this part of the sky.

X-ray clusters

Studies of x-ray emitting galaxy clusters – clusters which contain typically about six times as much hot x-ray emitting gas as optically luminous material – provide another means of testing cosmological models and getting a handle on the properties of dark energy. As we saw in Chapter 5 astronomers can use measurements of how the x-ray brightness and temperature of its extensive x-ray emitting gas cloud varies with distance from its centre to work out the total mass of the cluster (including baryonic and non-baryonic matter) and what fraction of that mass is contained in the hot gas (the gas fraction) or in the form of baryons (the baryon fraction). Because there is no known process that can have changed significantly the baryon fraction in huge structures such as these, astronomers usually assume that the baryon fraction and the gas fraction (the hot gas contains most of the baryons) in x-ray emitting clusters is constant and should, therefore, be the same in clusters at different redshifts.

In 1996, Japanese cosmologist Shin Sasaki and Harvard cosmologist Ue-Le Pen pointed out that the observed x-ray luminosity, temperature profiles and apparent angular diameters of remote clusters will only yield a

throughout the universe. This technique has great potential. Because gravitational distortions reveal the distribution of the dark stuff as well as the luminous stuff, weak lensing surveys have been described as offering an 'unrivalled "clean" tracer' of the dark matter distribution which is completely independent of the galaxy distribution surveys.

constant ratio of baryonic mass to total mass if the diameter distances (distances worked out from their apparent sizes) of those clusters are matched to the correct cosmological model. If the wrong values of cosmological parameters (such as matter density and dark energy density) are adopted, the measured baryon fraction will appear to change with increasing redshift. Sasaki and Pen suggested that measurements of this kind could be used to test cosmological models.

This challenge was taken up by Steven W Allen, of the Institute of Astronomy, Cambridge, UK. Together with colleagues from Germany and the USA, he used the Chandra X-ray Observatory (the world's most powerful orbiting x-ray telescope) to make a detailed study of 26 of the largest x-ray-emitting clusters in the redshift range 0.07 to 0.9, which reaches back to the time when the effects of dark energy began to become dominant and the expansion of the universe began to accelerate. The results, which were published in 2004, showed that the measured fraction appears to decrease with increasing redshift if the 'old' standard cold dark matter ($\Omega_{CDM} = 1$, $\Omega_\Lambda = 0$) were adopted, but becomes constant with ΛCDM models with dark energy included. When combined with data from other sources, Allen's x-ray cluster results yielded a value for the dark matter equation of state which, while consistent with $w = -1$ (the cosmological constant value) within the limits of error, nevertheless favoured a slightly more negative value (around -1.2) suggestive of phantom dark energy that increases with time.

Dimples in space

New and independent evidence for the space-stretching effects of dark energy emerged in late 2003 when four different observing teams succeeded in detecting a phenomenon known as the integrated Sachs-Wolfe (ISW) effect, a subtle modification of the pattern of warmer and cooler patches in the cosmic microwave background that had been predicted nearly 40 years earlier by Rainer Sachs and Art Wolfe. What happens is this: when a photon which is travelling towards us from the surface of

  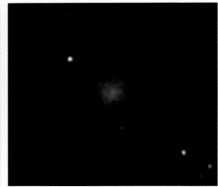

last scattering falls into the gravitational well created by a concentration of mass, it gains energy (like a ball rolling into a depression) and becomes blueshifted (shifted to a shorter wavelength). If the gravitational well does not change in any way while the photon is in transit, when the photon climbs back out of the well on the other side (like a ball rolling uphill) it will lose all the energy that it had gained and will be redshifted back to its original wavelength. But if, while the photon is travelling through a 'cosmic dimple' of this kind, the repulsive influence of dark energy has caused additional stretching of space, the dimple will be partially flattened out; the emerging photon will lose less energy than would otherwise be the case and will retain part of the blueshift that it gained on the way in.

Between leaving the surface of last scattering nearly 14 billion years ago and arriving at the Earth today, microwave background photons have passed through numerous denser and more rarefied regions of space. If matter were the only constituent of the universe, the redshifts and blueshifts would cancel out, but if repulsive dark energy is stretching and partially flattening the intervening cosmic dimples, additional marginally warmer (blueshifted) and marginally cooler (redshifted) patches will be superimposed on the microwave background temperature map. In the early stages of the evolution of the universe, matter would have been the dominant cosmic constituent, and the effects of dark energy would have been inconsequential. But at more recent times, the influence of dark energy would have become much more pronounced, giving rise

to perceptible ISW effects.

Although it is extremely hard to disentangle the ISW effect from primordial temperature fluctuations using microwave background data alone, the teams that reported its discovery in 2003 unveiled its subtle influence by correlating WMAP data with the large-scale distribution of foreground matter as revealed by surveys of galaxies, radio galaxies and x-ray data. With the improvements in sensitivity and resolution that are likely to be achieved over the next decade or so, the ISW effect is likely to become a powerful probe of dark energy and its equation of state.

Baryon oscillations – a new standard ruler

Whereas Type Ia supernovae are the preeminent standard candles for exploring the expansion history of the universe, a recently observed phenomenon called *baryon acoustic oscillations* promises to provide a valuable new standard ruler.

As we saw in Chapter 8, up until the time when matter and radiation decoupled from each other (some 380,000 years after the beginning of time) the competing influences of gravity and the pressure exerted by the closely coupled mix of baryons and photons set up a series of oscillations in this photon-baryon 'fluid'. In any region where the density was a little higher than average, gravity would attempt to pull the mix of ordinary matter (baryons and electrons), photons and cold dark matter together. However, whereas the cold dark matter particles, which did not interact with the photons, could fall together quite happily, the pressure exerted by the closely-coupled

The three panels show, from left to right, x-ray emitting clusters of galaxies, at progressively greater distances (and seen as they were 1 billion, 3.5 billion and 6.7 billion years ago, respectively) as imaged by the orbiting Chandra observatory. Measurements of the relative proportions of very hot gas and dark matter in clusters such as these can be used to measure distances and the composition and expansion history of the universe.

The origin of the 'ISW' effect is illustrated schematically here. Like a ball rolling in and out of a smooth hollow (a), a photon passing through a gravitational well (c) emerges with is wavelength unchanged. But like a ball rolling down a high slope and climbing a lower one, a photon will emerge with more energy and a shorter wavelength if the gravitational well becomes shallower while it is passing through.

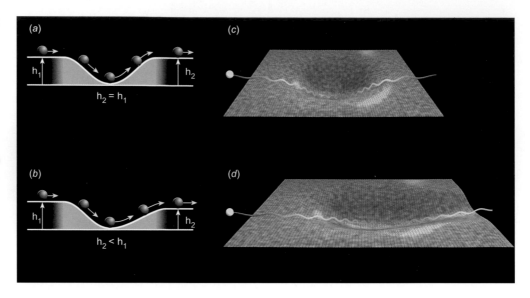

mix of baryons, electrons and photons halted the collapse of ordinary matter and caused it to rebound. This process caused regions containing more than average amounts of matter to oscillate in and out, sending out waves which propagated through the photon-baryon fluid in much the same sort of way as ripples spread across a pond when you throw in a pebble. These waves travelled outwards at the speed of sound, which, in the high-temperature photon-baryon fluid, was extremely high (a large fraction of the speed of light).

But when the temperature dropped low enough for atomic nuclei to capture electrons and create neutral atoms, and radiation and matter decoupled from each other, the speed of sound plummeted by an enormous factor. The acoustic ripples ceased to expand and were frozen into the distribution of baryonic matter at the size to which they had grown at the time of decoupling. The radii of the largest ripples were equal to the radius of the sound horizon – the maximum distance to which a sound wave could have travelled by the time the universe was about 380,000 years old. This is a figure that can be calculated with quite good precision from our knowledge of the density of matter and radiation, and the temperature, at that time. Stretched by the ongoing expansion of space, the fossilised imprints of these primordial ripples should by now be about 500 million light-years across.

In January 2005, a team of SDSS researchers announced that by analysing the distribution of some 46,000 very luminous red galaxies, at a mean redshift of 0.35, they had found a slight excess of galaxies with separations of 500 million light-years – exactly the predicted signature of sound waves that had been frozen into the distribution of baryons at the time of decoupling. A similar independent analysis on a different dataset by the 2dFGRS team, which was published at more or less exactly the same time, also found the imprint of primordial sound waves. These detections confirm two key predictions of cold dark matter models: that features such as the early acoustic ripples are preserved as the expansion continues, and that density fluctuations evolve under the influence of gravity in proportion to the expansion factor of the universe. The results also provide further proof that cold dark matter was the dominant form of matter at the time of decoupling for, had most of the matter been baryonic, the fossilised density ripples would have been much more prominent.

Cosmologists are excited by the prospect of using acoustic oscillations as 'standard rulers' for measuring the expansion history of the universe. Because the physical size of the baryon oscillations is known, then if astronomers can measure the angular diameter of its imprint in the distribution of galaxies at some particular value of redshift, they can work out its diameter distance. If they could measure its angular diameter at a range of different redshifts, they would be able to plot a 'Hubble diagram' that relates diameter distance to redshift. This would give them a way of charting the expansion history of the universe which is analogous to, but completely independent from, the luminosity distance-redshift relationship provided by the supernova data.

Astronomers have already taken the first step along that road. Following the discovery of the acoustic feature in the distribution of galaxies, the angular size of this 'ruler' has now been measured at two very different redshifts. The SDSS team measured it at a redshift of 0.35, and from that measurement were able to calculate, to an accuracy of about 5 percent, that the actual distance from the Earth to that particular redshift is about 1370 megaparsecs (4470 million light-years). The other measurement comes from the microwave background. The measured sizes of the most prominent warm and cool patches in the microwave background, which are the imprint of acoustic oscillations on the last scattering surface, gives the angular size of the ruler at a redshift (according to WMAP) of 1089. By comparing these two measurements of the angular size of the ruler, the SDSS team were able to determine the ratio of the distances to these two redshifts and calculate the amount by which

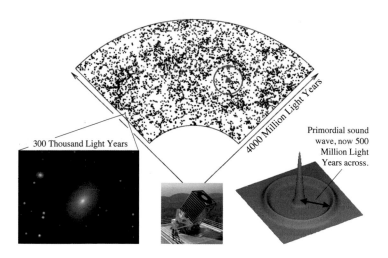

300 Thousand Light Years

4000 Million Light Years

Primordial sound wave, now 500 Million Light Years across.

the universe has expanded since the microwave background radiation was released.

However, baryon acoustic oscillations have more to offer than diameter distance alone. Diameter distance relies on measuring the angular diameter of the acoustic imprint on the plane of the sky but, because these features show up in three dimensions in galaxy redshift surveys, astronomers can also pick out their imprints along the line of sight (in the radial direction), and can therefore measure the *difference* between the redshifts of galaxies on the far side and the redshifts of galaxies on the near side of each imprinted ripple. Astronomers can use this information

A map of the galaxies (dots) in a portion of the Sloan Digital Sky Survey (SDSS) is shown at the top. The position of the Earth is at the bottom, represented by a picture of the SDSS telescope. The bull's eye shows the present-day scale of the imprint of primordial sound waves; however, the imprint is too subtle to see by eye.

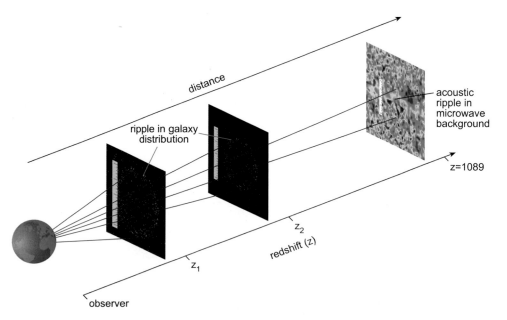

distance

ripple in galaxy distribution

acoustic ripple in microwave background

z_2

$z=1089$

z_1

redshift (z)

observer

The size of the fossilised imprints of primordial baryon oscillations, which show up in the microwave background and, very subtly, in the distribution of galaxies at different redshifts, can be used as a 'standard ruler' to measure distances.

to calculate the value of the Hubble parameter (the Hubble 'constant') at that redshift. Consequently, the baryon oscillation technique has the great advantage that it allows astronomers to measure both the distance to a particular redshift *and* the Hubble expansion rate at that redshift. By putting these two things together, and repeating the exercise for acoustic ripples at a range of different redshifts, cosmologists should eventually be able to map out how the speed of cosmic expansion has changed over the history of the universe.

Another advantage of this technique is that it is unaffected by some of the difficulties that afflict Type Ia supernova measurements – the extinction of light by dust, the possible evolution of supernova properties over cosmic time, and the influence of gravitational lensing on their perceived brightness. The disadvantage is that, because the imprints of baryon oscillations on galaxy distributions are extremely weak and extend across such large distances, astronomers will need to survey huge volumes of space in order to pick out the effect sufficiently clearly, and over a large enough range of redshift to provide useful constraints on the cosmological parameters.

The baryon oscillation method provides a potentially fruitful way of measuring the changing influence of dark energy on the expansion history of the universe and thereby determining its equation of state, w, and whether, and if so how, it changes with time. Plans are already afoot to conduct large-scale surveys of this kind with the extended SDSS II and a whole range of medium- and longer-term survey projects, some of which are described later in this chapter. Many of these surveys will rely on using relatively crude photometric redshifts (as described on page 161), the advantage of this approach being that vast numbers of galaxies can be imaged simultaneously and basic colour information can be acquired using much shorter exposure times than those which would be needed to obtain detailed spectra. The disadvantage is that, without detailed spectra, astronomers cannot obtain *precise* redshift values.

With that in mind, an international team of astronomers from the US, UK, Australia, Canada and Japan is proposing to construct a major new instrument called the Wide-Field Multi-Object Spectrograph (WFMOS), which will be capable of measuring spectra for up to 5,000 objects at a time when used on the 8-metre Gemini and Subaru telescopes. The aim is to conduct a survey of some 2 million galaxies at redshifts of less than 1.3, and of half a million galaxies at redshifts of between 2.5 and 3.5, in order to identify the imprint of baryon oscillations, and thereby probe the relationship between distance, expansion rate and redshift with accuracies in the region of 2 percent. Hopefully, the system will be up and running by around 2012.

Wringing more detail out of the microwave background

The cosmic microwave background holds a rich treasure trove of cosmological data which cosmologists are determined to exploit to the full. At ground level, projects such as the Very Small Array (VSA) on Mount Teide in the Canary Islands, and the Cosmic Background Imager (CBI), located at a height of more than 5,000 metres in the Chilean Andes, and several others, are continuing to provide high quality data. Meanwhile, at its lonely outpost in space, some 1.5 million kilometres from Earth, the WMAP probe is expected to continue returning data until around 2009. Analysis of the first year's worth of data from this legendary instrument produced a phenomenal wealth of high-quality cosmological information. The eagerly awaited second data release, based on the first three years' worth of WMAP measurements, took place in March 2006. With more than three times as much data to work with, and using improved analytical techniques, the WMAP Science Team has been able to achieve substantial improvements in the precision of their microwave sky maps and in the measured values of key cosmological parameters. Whereas the first WMAP data release concentrated primarily on producing high resolution maps of the warmer and cooler patches, the second focused, in addition, on the much more difficult task of measuring polarisation signals which, as we saw in Chapter 8, are produced when light waves scatter off charged particles such as electrons.

The oscillations that were taking place in the primordial mix of matter and radiation up until the time of decoupling consisted of expanding and contracting clumps of baryons and electrons. Because photons would have rebounded with slightly more energy when they scattered off expanding blobs and slightly less energy when they scattered off contracting blobs (i.e. they would have been Doppler shifted to shorter and longer wavelengths), patches of apparently warmer and cooler radiation will also appear in maps of *polarised* microwave background radiation. These can be plotted against angular size to produce a graph, or spectrum (the angular power spectrum of polarised radiation), the peaks and troughs of which will hold additional information relating to the mix of baryons, cold dark matter, neutrinos and dark energy that existed at the time when the microwave background radiation was released. But whereas the peaks in the spectrum of hotter and cooler patches revealed by ordinary (total intensity) temperature maps correspond to regions of maximum and minimum compression, the peaks in the primordial polarisation spectrum correspond to where their rates of expansion and contraction are greatest, which is mid-way between maximum and minimum compression. Consequently, the peaks in the polarised spectrum should coincide with the troughs in the spectrum of the ordinary temperature maps.

The detection of the faint imprint of polarisation in the cosmic microwave background was reported in 2002 by an American team which had used the Degree Angular Scale Interferometer (DASI), a instrument that had been set up at the South Pole. Since then, WMAP, CBI and the BOOMERanG-03 balloon-borne experiment, which flew in 2003, have also detected polarisation in the microwave background. The results show that the peaks and troughs in the spectrum of polarised emission are shifted in phase by half a cycle compared to the peaks and troughs in the total intensity (ordinary temperature map)

The Very Small Array (VSA), located on Mt. Teide, Tenerife, consists of 14 small microwave receivers, each of which has an antenna only 15 centimetres in diameter. As shown here, each antenna has a larger feed horn.

spectrum, exactly as theory predicts. This, of itself, provides striking confirmation that the fluctuations in the microwave background are indeed caused by acoustic oscillations in the primordial mix of matter and radiation.

As we have already seen, one of the major difficulties facing astronomers who are attempting to interpret the temperature and polarisation maps is how to disentangle temperature fluctuations (which were imprinted as cosmic fossils on the surface of last scattering at the time when matter and radiation decoupled from each other) from the various 'foreground' effects. These effects are caused by scattering of microwave background photons as they pass through clouds of ionised gas in galaxy clusters and in the Milky Way Galaxy itself, by phenomena such as gravitational lensing and by the formation of the first stars. All of these phenomena plant their own fingerprints

This detailed all-sky picture of the infant universe was constructed from three years of WMAP data. The image reveals 13.7 billion year-old temperature fluctuations (shown as colour differences), and spans a temperature range of ±200 microkelvin.

Peaks in the predicted angular spectrum of polarised microwave background radiation (green line) line up with troughs in the total intensity spectrum of the hot and cool patches that are seen in conventional microwave temperature maps (red line). The polarised signal is a hundred times weaker than the total signal, and has been exaggerated for clarity. The Cosmic Background Imager polarisation results fit the predicted curve.

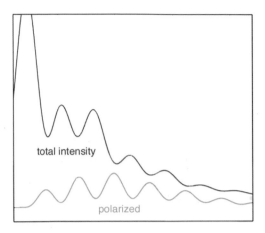

on the microwave background.

When the first massive stars began to form, they ionised the surrounding clouds of gas, so releasing large numbers of free electrons which created a kind of cosmic fog through which background photons had to travel. Scattering of light by this cosmic fog reduced the observed amplitudes (relative brightnesses) of the hot and cool patches and also imprinted patterns of polarisation (preferred directions of vibration) on the microwave radiation. The three-year WMAP data have revealed detailed patterns of polarisation across the whole sky for the first time (previous polarisation measurements by ground-based and balloon-borne experiments only revealed the distribution of small-scale features in selected regions of the sky; by mapping the whole sky, WMAP also uncovered large-scale patterns of brightness and polarisation). By observing and mapping the sky at five different wavelengths, WMAP researchers have been able to construct a model of the foreground emission and polarisation caused by the Milky Way and its environs, which they have subtracted from the observed microwave maps to 'clean them up'. Using their all-sky measurements, which detected weak polarisation imprints that were

The white bars superimposed on the WMAP all-sky temperature map (based on three years of data) show the polarisation direction of the oldest light in the universe. This new information helps to pinpoint when the first stars formed and provides new clues about events that happened a microscopic fraction of a second after the Big Bang.

more than a hundred times feebler than the brightness differences in the all-sky temperature maps, the WMAP experimenters were able 'cleanly' to separate out effects caused by the re-ionisation of gas by the first stars from genuine imprints of the density fluctuations that were present at the time of decoupling.

The three-year WMAP data indicate that the first stars switched on when the universe was about 400 million years old (which corresponds to looking back to a redshift in the region of $z = 11$). The pattern of hot and cool spots left over when the various foreground contaminations were removed (as best as possible) showed that the brightnesses (or 'amplitudes') of the smaller patches were slightly lower than the brightnesses of larger ones, much as long-wavelength ocean swells tend to be higher (have a greater amplitude) than shorter wavelength waves and ripples. This (fairly gentle) decrease in brightness with decreasing angular size is measured by a quantity called the 'scalar spectral index' (n), which, according to the 3-year WMAP data, works out to be 0.95, with a possible error of about 2 percent either way. This is a result that matches extremely well with predictions made by the simplest inflation models but which contradicts a prediction associated with the simplest model of how structure formed in the universe, that the brightness would be the same for features of all sizes (i.e., n = 1). The new WMAP results, therefore, provide further powerful support for straightforward inflationary theory and appear to go a long way towards eliminating a number of alternative theories, and some of the more complex ideas about the nature of inflation.

The three-year WMAP data have also enabled researchers to refine their estimates of the relative proportions of baryonic matter ('atoms'), non-baryonic dark matter and dark energy. The proportions emerging from the new data are, in round figures: baryons 4 percent ($\Omega_b = 0.04$); non-baryonic cold dark matter, 22 percent ($\Omega_{CDM} = 0.22$); and dark energy 74 percent ($\Omega_\Lambda = 0.74$). These figures are similar to, but more precise than, those which had been calculated in 2003 using just one year's worth of data.

Crucially, the improved data has enabled WMAP Science Team members to home in more closely on the equation of state of dark energy - the all-important w-value. If it is assumed that the universe is precisely geometrically flat and that w is constant, the new WMAP data when combined with recent new supernova data (from the 'SNLS' programme described later in this chapter) give a w-value of –0.97 with a probable error of less than 10 percent either way. If instead the value $w = -1$ (the cosmological constant value) is assumed, the observations show that the overall mean density of the universe is equal to the critical density to within about 1–2 percent. Even if the assumption of flatness is set aside, the new WMAP data, when combined with large scale structure measurements and supernova results, produce a value, $w = -1.06$, with a probable error of around 10 percent. In essence, the new WMAP results confirm that the value of w is very close to –1 (the value that matches the cosmological constant/vacuum energy interpretation of dark energy), give or take about 10 percent at most.

Initial indications suggest that the new temperature maps still contain (or still appear to contain) what the WMAP Science Team itself describes as 'questionable features'. Some would regard these as evidence for a preferred axis of alignment among the large-scale hot and cold spots, and discernable differences between the patterns of hot and cold spots on opposite hemispheres of the sky, and an unusually deep and large cold spot in the southern sky - features which require further investigation and explanation. But leaving aside these as yet unexplained anomalies, the new three-year WMAP results reinforce, improve and extend what the one-year results had shown, and give further strong support for, and confidence in, the inflationary ΛCDM model. Indeed, in one of their first papers on the 3-year data[2], the WMAP Science Team remark, 'The standard model of cosmology has survived another rigorous set of tests', and go on to say 'The data are now so constraining that there is little room for significant modifications of the basic

ΛCDM model' – a ringing endorsement of the status quo.

The next leap forward in studies of the microwave background is expected to come from the European Space Agency's Planck spacecraft, which is scheduled to be launched in 2007. Equipped with a 1.5-metre antenna that focuses radiation on to two arrays of detectors, it has been designed to detect temperature differences of a few millionths of a degree at angular resolutions down to about a sixth of a degree – a factor of two improvement on WMAP's capabilities. Like WMAP, it will be stationed at L2, 1.5 million kilometres away on the opposite side of the Earth from the Sun.

Towards the next decade

Looking ahead to the next decade, cosmologists have proposed a range of ambitious ground-based and space-based projects to confront key cosmological issues head on. One of the most exciting contenders is SNAP (the SuperNova/Acceleration Probe), a satellite mission concept that has been developed over the past several years by a team headed by Saul Perlmutter, of the Supernova Cosmology Project. The proposed spacecraft would carry a 2-metre wide-angle optical and infrared telescope equipped with a half-billion pixel camera and a spectroscopic system. Its primary aim is to measure redshifts and brightness for around 2000 Type Ia supernovae across the whole range of redshifts from $z = 0.1$ to $z = 1.7$; in addition, it would undertake a weak lensing survey of subtle galaxy distortions, spanning around 1000 square degrees of sky. With the combination of supernova and weak lensing data, SNAP should be capable of measuring the expansion history and matter density of the universe to an accuracy of about 1 percent, the present dark energy equation of state (w) to around 5–10 percent and its variation over time to within 10–20 percent, which should be good enough to eliminate at least some of the competing models of dark energy.

SNAP is one of the candidates for the Joint Dark Energy Mission (JDEM), which is to be funded by NASA and the US Department of Energy, with a launch date possibly as early

An artist's impression of the Planck spacecraft showing the telescope, which consists of two mirrors encased by a shield, at the top. The thicker base is the service module.

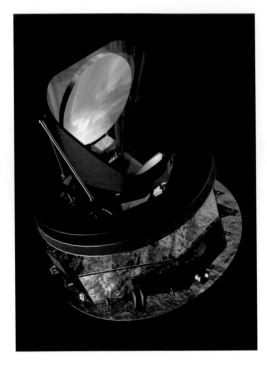

An artist's cutaway impression of the proposed SNAP (SuperNova/Acceleration Probe) spacecraft and its optical system.

as 2011, but more likely a year or two later. Another contender is the delightfully named (at least for 'Star Wars' enthusiasts) JEDI (Joint Efficient Dark-energy Investigation) mission. This more elaborate proposal, which also envisages using a space-based 2-metre telescope located at L2, would utilise Type Ia supernovae as standard candles and baryon oscillations as standard rulers to chart the expansion history of the universe. It also would map weak lensing at different redshifts to search for departures from the standard ΛCDM cosmology. The aim is to measure at least 14,000 Type Ia supernovae out to a redshift of 1.7, obtain redshifts for about 100 million galaxies in a 10,000 square degree patch of sky out to $z = 2$, and redshifts for at least 10 million galaxies in a 1000 square degree region of sky to a maximum redshift of 4. Provided that other experiments (such as Planck) have been able by that time to measure the density of matter to an accuracy of 1 percent, JEDI ought to be able to determine whether the dark energy equation of state deviates from the cosmological constant value ($w = -1$) by as little as 1–2 percent, and to measure its rate of change to an accuracy of around 5 percent.

Back on the ground, one of the most ambi-

tious projects for the next decade is the Large Synoptic Survey Telescope (LSST). As currently envisaged, this instrument will have a unique design, based around an 8.4-metre concave primary mirror and a 4-metre convex secondary mirror (itself as big as the main mirrors on many large research telescopes), which will enable it to image an area of sky some 50 times larger than the apparent area of the full Moon every 15 seconds. Equipped with a 3 billion pixel digital camera and a range of colour filters, it will repeatedly survey some 30,000 square degrees of sky and measure the shapes, brightness and colours of billions of galaxies. Its primary goal is to carry out a comprehensive weak lensing study at a range of different redshifts, in order to produce a three-dimensional map of the spatial distribution of mass extending back to when the universe was about half its present age. Astronomers expect that it will detect, weigh and locate in three dimensions hundred of thousands of mass concentrations and, from this data, precisely delineate the evolution of structure in the universe and the influence of dark energy on that process.

In addition to undertaking a host of other tasks (including looking for potentially hazardous near-Earth asteroids), LSST will also have the potential to detect thousands of Type Ia supernovae a year, out to redshifts in the region of $z = 2$. By combining weak lensing and supernova data, and utilising precision measurements of the cosmic matter density from WMAP and the forthcoming Planck mis-

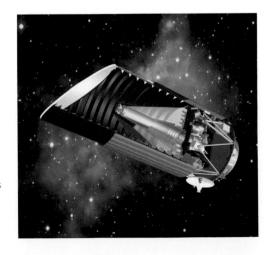

sion, LSST should be able to pin down the value of the dark energy equation of state to within around 1 percent and its rate of change to within about 5 percent. These are performance figures that are comparable to those which are anticipated for SNAP or JEDI.

Space-based and ground-based projects each have their strengths and weaknesses. Space missions operate clear of the Earth's atmosphere and can therefore obtain stable images, free from the effects of atmospheric turbulence or background sky brightness, 24 hours a day. Crucially, they can observe unimpeded in the near-infrared region of the spectrum. This is of vital importance when looking at high-redshift supernovae because, at redshifts greater than about $z = 1$, most of their light is shifted into the near-infrared region of the spectrum which is largely inaccessible from the ground. The disadvantages are high cost, inaccessibility (if anything goes wrong no one can get out to L2 to do anything about it) and the sheer technical problems involved in relaying immense amounts of data back to Earth. The much larger ground-based telescopes, such as LSST, can detect faint objects with much shorter exposure times, are cheaper to build (although still expensive), can readily be maintained and modified, do not have to transmit data back across space, but do have to peer out through the atmosphere.

Both types of project – ground-based and space-based – have an immense amount to offer, would complement each other extremely well, and would provide vital cross-checks on each other's results. The LSST project has already been allocated substantial development funding, and should be up and running by 2012. Hopefully SNAP, or JEDI or some other version of the Joint Dark Energy Mission will be operating in space by about the same time, or within a few years thereafter.

In the shorter term
In the meantime, there is a rich abundance of observational projects – not least the ongoing SDSS – currently under way, or projected to be up and running within the next few years.

Astronomers are working hard better to

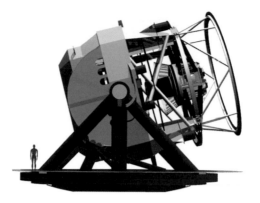

An impression of the design, current as of April 2006, for the 8.4 metre Large Synoptic Survey Telescope (LSST), seen with a human figure to give scale. It will use a special three-mirror system to create an exceptionally wide field of view.

characterise the properties of Type Ia supernovae, by observing as many relatively nearby supernovae as possible – supernovae that can be studied in greater detail than the more remote ones – through programmes such as the 'Nearby Supernova Factory', which utilises images acquired by a telescope that is dedicated to searching for near-Earth asteroids, and the Lick Observatory Supernova Search, which uses an automated robotic telescope to hunt down its quarry. The accumulated data will allow astronomers to compile better 'templates' of Type Ia spectra and light curves, which will make it easier to measure the peak luminosities of high-redshift supernovae and check on the extent to which their properties have evolved over the history of the universe. With improved reference spectra, astronomers should be able to make better allowance for the effects of dust extinction in the host

In order to simulate the appearance of a space-time warp, the detailed mass distribution in the galaxy cluster CL0024 is shown as if a large sheet of graph paper were located behind this particular gravitational lens. In addition to measuring strong gravitational lensing effects like this, the LSST will measure more subtle distortions caused by weak lensing (cosmic shear).

galaxies, improve their chances of picking out the effects (if any) of hypothetical grey dust, and whittle down errors of observation and interpretation.

One of the current projects is ESSENCE (Equation of State: SupErNovae trace Cosmic Expansion – a wonderfully outlandish acronym!), the goal of which is to use a sample of about 200 Type Ia supernovae at moderate redshifts ($0.2 \leq z \leq 0.8$) to place constraints on the equation of state of the universe. It uses spectroscopy, not only to obtain the redshifts of the supernovae but also to confirm that they have been correctly identified as Type Ia (whereas the widely-used technique of plotting the light curve of a distant supernova and comparing it with a series of templates is very effective, the only way to be absolutely sure that an object is a Type Ia supernova is through spectroscopy). The ESSENCE survey uses a wide-field CCD camera called MOSAIC, attached to the 'Blanco' 4-metre telescope at the Cerro Tololo Inter-American Observatory (CTIO) in Chile, to identify potential supernovae, which are then studied spectroscopically by larger and more powerful instruments. The programme, which started in September 2002 and is scheduled to continue until 2007, has already established that high-redshift Type Ia supernovae show strong similarities to low-redshift ones, which implies that no significant evolutionary changes appear to be taking place. This bolsters their credentials as standard candles.

An international collaborative programme called the CFHT Legacy Survey, which is led by French and Canadian observers, is using a wide-field CCD Camera called Megacam attached to the 3.6-metre Canada-France-Hawaii Telescope on top of Mauna Kea, Hawaii, to conduct a 'wide' survey of some 170 square degrees of sky and a 'deep' survey of 4 square degrees of sky. Their aims are: (1) to use weak lensing and galaxy distributions to map the evolving distribution of mass structures in the universe; and (2) to detect and monitor up to 2000 Type Ia supernovae. The survey, which began in 2003, is scheduled to run for about five years. The supernova part of the programme, which is called the Supernova

Legacy Survey (SNLS), uses the CFHT camera to detect and monitor supernovae and makes follow-up spectroscopic observations with some of the world's largest telescopes in order to measure their redshifts. By combining their first year's crop of supernova observations with the Sloan Digital Sky Survey's measurement of baryon acoustic oscillations (described earlier) the SNLS collaboration had by the end of 2005 already produced measurements of cosmic matter density and the dark energy equation of state with a precision of around 10 percent. By the time the survey has been completed, the precision of those measurements is expected to double or even treble.

In a similar, although more comprehensive, vein, the Dark Energy Survey (DES) is another five-year project, which is currently being developed by a US-led international consortium. The intention is to study the nature of dark energy and cosmic acceleration through two coupled surveys – a scan of 5000 square degrees of sky designed to produce redshift data for some 300 million galaxies in the range $z = 0.2$ to $z = 1.3$; and a 40 square degree search for supernovae. The Dark Energy Survey will study the expansion history of the universe and the growth of structure using four distinct techniques: (1) a survey of the numbers and distribution of galaxy clusters out to redshift 1.3; (2) a weak lensing study of the distribution of dark matter in several different redshift shells; (3) a study of how the angular distribution of galaxies varies with redshift (the galaxy angular power spectrum); (4) a supernova search that is expected to yield around 2000 Type Ia supernovae with redshifts between 0.3 and 0.8 over the five-year lifetime of the programme. Taken together, and combined with WMAP or Planck data, these techniques should be able to deliver a figure for w that has an accuracy of about 2–3 percent if w is constant, or 10–20 percent for its variation, if w changes with time.

The state of the universe

Over the course of the next five to ten years, projects such as these – and there are many more under way or planned than I have mentioned here – will provide a wealth of data to test, challenge, confirm or possibly refute the key features of the Standard Model of the universe.

By subjecting the microwave background to even deeper and more detailed scrutiny, cosmologists should be able to test whether the predictions of inflationary theory are valid. Is space truly flat, or is there some slight residue of curvature there? Do the primordial density fluctuations whose signatures are imprinted on the microwave background really match what inflationary theory predicts, or does reality deviate from the theoreticians' ideal? Does the apparent detection of non-random patterns in sizes, shapes, orientations and distribution of some of the spots (particularly the cooler spots) in the microwave background point towards instrumental errors, the influence of foreground phenomena, or towards issues that could undermine the whole inflationary model?

By mapping more precisely and in much more detail the way in which matter – both luminous and dark – is distributed, and how its distribution has changed over cosmic time, cosmologists will be able to see more clearly

exactly how density fluctuations have grown and evolved over the history of the universe to create the rich pattern of galaxies, clusters, filaments, walls and voids that we see in the present-day universe. They will also be better able to see how well, or otherwise, the growth history of structure matches with the predictions of cosmological models. By combining observations of the evolution of cosmic structure with improved microwave background data, they will be able to get a better handle on the relative proportions of the various cosmic ingredients – baryons, cold dark matter particles, neutrinos, dark energy, and anything else that can be factored into the mix.

With access to a hugely increased sample of Type Ia supernova data of higher quality across a greater range of redshifts, they will be able to confront potential problems such as the effects of dust and evolution, and chart out the entire expansion history, over the past 10 billion years or more, with great precision. They will be able to cross-check that history with other techniques, for example by using x-ray emitting clusters and baryon oscillations as standard rulers to map out the relationship between angular size and redshift. With a better knowledge of when the change from decelerating to accelerating expansion took place, and improved measurements of early deceleration and late acceleration, they will be able directly to see the influence of dark energy in the expansion history of the universe and measure its repulsive negative pressure.

Cosmologists will close in ever more tightly on the values of the key cosmological parameters, even though, should the mean overall density of the universe actually be *precisely* equal to the critical density ($\Omega_{total} = 1$) we will never be able to know that for sure because, however good our measurements may become, there will always be a tiny margin of uncertainty. Nevertheless, whereas the present uncertainties in the various density parameters – for matter as a whole (Ω_M), baryons (Ω_b), cold dark matter (Ω_{CDM}), neutrinos (Ω_v), dark energy (Ω_{DE}) – range from a few percent to around 10 percent, within a decade they may well have been whittled down to around 1

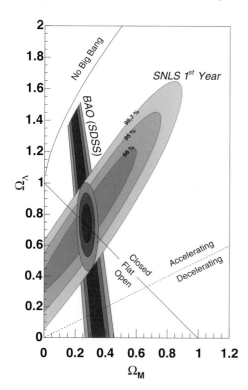

The oval red contours show different levels of confidence in the values of matter density (Ω_M) and dark energy density (Ω_Λ) obtained by combining SNLS supernova data (green ovals) and SDSS measurements of baryon acoustic oscillations (blue contours). When the assumption that space is flat is taken into account, the data indicate values in the region of $\Omega_M = 0.27$ and $\Omega_\Lambda = 0.73$.

percent, in which case we will have a pretty impressive inventory of the overall composition of the cosmos.

But the big mysteries remain. We do not know the nature of dark matter particles, though there are plenty of potential candidates and immense amounts of effort are going into searching for them (although if MOND is valid – see Chapter 6 – they may not even exist at all). But within the decade, they may have been created in particle accelerators such as the Large Hadron Collider, detected directly in underground laboratories, or revealed through their telltale signatures in astrophysical or atmospheric phenomena. But the biggest mystery of all is, what is the nature of the enigmatic dark energy which dominates the universe today, and which seems to have gained the ascendancy just a few billion years ago? Is it Einstein's cosmological constant, Λ, in its modern guise of constant density vacuum energy? Is it something dynamic and changing like quintessence? Is it phantom dark energy, which will eventually escalate out of control and tear the universe apart in a terminal Big Rip? Or will it eventually fade away, or reverse its sign, thereby condemning the universe to collapse into a Big Crunch? Does it fluctuate periodically; is there more than one kind of dark energy? Are dark energy and dark matter two sides of the same coin?

Measuring the ratio of pressure to density (the equation of state, w), and determining whether or not it varies with time, is central

to cosmologists' attempts to unveil the nature of dark energy or, if nothing else, to eliminate many of the burgeoning range of theoretical possibilities. As yet, the data are not good enough to pin down w with sufficient accuracy to distinguish between vacuum energy, quintessence, phantom models, and the rest. Measured values, derived from a variety of different kinds of observations, range between around –0.6 and –1.9, with some supernova and cluster x-ray measurements tending to favour values which are *more* negative than –1, and WMAP data pointing towards values which are equal to, or marginally *less* negative than –1.

Most of the presently determined values, particularly those which have been computed by combining various different sets of data, are sufficiently close to –1, within their margins of error, to be consistent with the cosmological constant (vacuum energy). For the moment, Einstein's Λ matches the evidence. It remains to be seen whether or not the huge improvements that the next decade hopefully will bring will lead experimenters to a distinctly different value. The worrying possibility is that if dark energy has an equation of state very close to –1, and changes only very slowly with time, cosmologists could find themselves in the frustrating position, despite their best efforts, of never being able to distinguish the equation of state (and hence the nature of dark energy) from what has come to be known as 'vanilla' cosmology – the rather bland concordance cosmology with a cosmological constant (ΛCDM).

The subtle effects that may eventually allow us to distinguish between different forms of dark energy lie at, or even beyond, the limits of present-day observational techniques. With each advance, we probe that much deeper but at the same time unveil new puzzles and challenges. What we most want to measure seems always to lie at or beyond the current limits, dimly glimpsed through the noise of uncertainty and error. It was ever thus. As Hubble remarked[3] in his book, *The Realm of the Nebulae*, '… we measure shadows, and we search among ghostly errors of measurement for landmarks that are scarcely more substantial. The search will continue.' Indeed, it will.

The oval red contours show different levels of confidence in the values of matter density (Ω_M) and the all-important ratio of dark energy pressure to dark energy density (w) obtained by combining SNLS supernova data (green contours) and SDSS measurements of baryon acoustic oscillations (blue contours). The resulting value of w is very close to, but possibly slightly more negative than, -1.

A Good Time to be Here

It's an exciting time for cosmology and particle physics. If mainstream cosmology has got the picture right, then the following statements are a fair summary of the constitution of the universe: (1) Baryons – the stuff of which atoms, planets, stars and galaxies are made – are a relatively minor constituent of the cosmos, contributing less than 5 percent to its total mass-energy budget. (2) Non-baryonic cold dark matter appears to outweigh the baryonic stuff by a factor of five, or thereabouts, but the overall mean density of matter in all its guises (luminous and dark, baryonic and non-baryonic) still adds up to less than 30 percent of the total mass-energy density of the universe. (3) Dark energy, the mysterious extra ingredient which is driving the current phase of accelerating cosmic expansion, is the dominant component, and contributes at least 70 percent of the total mass-energy budget. Although cosmologists and particle physicists have no shortage of theories and hypotheses about what dark matter and dark energy may be, as of now we still have to admit that we simply do not know the true nature of 95 percent of the contents of the universe.

The 'standard' inflation + cold dark matter + dark energy scenario seems to be backed up by a wealth of observational evidence, theory and simulations, and is an impressive edifice of intertwining physics that links the largest cosmic scales to the microcosm of particle physics. Yet not everyone finds the mainstream picture quite so compelling. There are some who feel that the standard Big Bang picture has become too complex and contrived, with a succession of add-ons and 'fixes' such as inflation, dark matter and dark energy brought in to hold it all together. This prompts parallels with the complex array of epicycles and related constructs which the second-century Greek astronomer and philosopher, Claudius Ptolemaeus (better known as Ptolemy), was forced to add to his theory of planetary motion in order to preserve the then deep-rooted paradigm of an Earth-centred universe and uniform circular motion in the heavens.

The standard paradigm is in the firing line. There are alternative approaches, notably Modified Newtonian Dynamics (MOND), which may be able to pose a genuine challenge to dark matter and dark energy. Should standard cosmology fail to meet some of the challenges with which it is confronted, then we may be forced to rethink parts, or *in extremis* all, of the widely-favoured inflationary Big Bang scenario; and that would make for interesting times.

As yet, no-one has actually detected any of the proposed cold dark matter particles in any experiment, although there are tantalising hints contained in gamma-ray emissions from the galactic centre and around the sky as a whole, and in the hotly disputed annular modulations reported by the DAMA team. Particle physicists may soon be able to create and detect some of the candidate particles in new generations of ultra-powerful particle accelerators such as the soon-to-be-commissioned Large Hadron Collider. With subterranean detectors likely to achieve orders of magnitude increases in sensitivity over the course of the next 10 years, there is every chance that galactic WIMPs may be detected within that timescale. But if WIMPs have still not been found a decade or so from now, doubts will surely grow about their very existence.

Observational cosmology is advancing by

leaps and bounds. From a discipline that used to be long on theory but exceedingly short on fact, cosmology is moving confidently into an era of precision measurement and, before too long, should be able to pin down key cosmological parameters, and the expansion history of the universe, with accuracies in the region of 1 or 2 percent. Over the next decade or so, cosmologists are confident of being able to measure the effects, properties and evolution (if any) of dark energy with sufficient precision, if not to decide exactly what dark energy *is*, at least to be able eliminate many of the burgeoning multiplicity of possibilities. A deluge of data is set to come our way, out of which a better, deeper and more complete understanding of the cosmos surely will emerge.

So, it's good to be around when cosmology and particle physics are poised on the brink of discoveries, which on the one hand may uncover key pieces of the cosmic jigsaw that will lead us towards a more coherent whole, but on the other, could undermine the existing edifice and open up possibilities as yet unimagined.

However, it's a good time to be around for wider and deeper reasons, too. We, as a species, exist at an interesting time, fairly soon (in cosmic terms) after the onset of the current, and perhaps eternal, phase of accelerating expansion. Had dark energy become dominant too early, matter would have been blasted apart too quickly to allow stars, planets and galaxies to form, and we should not have been here. Had dark energy been feebler, its effects would not yet have kicked in, we would not be aware of its presence, and the universe would have appeared, in a sense, 'less interesting' than it does today. Had we not arrived on the scene until many billions of years hence, the transition from matter to dark energy domination would have been much further back in time and evidence of the earlier deceleration phase might have been much harder to acquire. Whether our being here at this particular stage in the history of the universe is: the result of sheer coincidence; the logical outcome of some fundamental link between the nature of dark energy and the onset of structure forma-

tion; the result of anthropic 'selection' – our universe being one of the very few, out of a great ensemble of separate universes that has the right mix of properties to enable sentient life to exist; merely a situation which may be 'not *too* surprising' if the total lifetime of the universe turns out to be no more than a few times its present age; or 'none of the above', it's intensely stimulating to be able to have the debate.

And it's a good time, too, because we can see such a rich variety of structure around us – galaxies and clusters which are still growing, merging, evolving and, in the process, triggering new bouts of star formation. It won't be so forever. The boundary of the observable universe is set by the distance that light can have travelled since the Big Bang – currently 13–14 billion light-years. Although that horizon will continue to grow at the speed of light, the accelerating expansion will inexorably carry more and more galaxies over that horizon to be lost to our ken. The galaxies themselves will fade and die as their constituent stars run out of fuel and the reserves of gas from which to make new ones become exhausted. Unless the Big Rip or the Big Crunch brings the cosmos to an abrupt end before the demise of all the stars and galaxies, the long-term future appears bleak. If the expansion goes on forever without new phase changes which could bring the current bout of acceleration to an end – and perhaps generate new and unimagined forms of matter and structure – the present epoch of stars, galaxies and light will turn out to be no more than a infinitesimal blip in the history of a universe that is heading inexorably towards dark emptiness and an eternal 'Big Chill'.

We live in a universe that is dominated by its dark side. The rich variety of galaxies and cosmic structure will not be around forever, but it's good to be here to appreciate, enjoy, and attempt to understand it all.

Endnotes

Chapter 4

[1] N W Evans and V Belokurov, 'RIP: The MACHO Era (1974–2004)', in *Proceedings of the Fifth International Workshop on the Identification of Dark Matter*, edited by Neil J C Spooner and Vitaly Kudryavstev, pp. 141–150, World Scientific, 2005.

Chapter 5

[1] This assertion is in marked disagreement with results obtained from detailed studies of the cosmic microwave background radiation, which appear to rule out the possibility that warm dark matter could be a significant component of the universe; see Chapter 8.

Chapter 6

[1] Robert H Sanders, of the Kapteyn Astronomical Institute in the Netherlands has pointed out that the adopted value of the transition acceleration, a_0, is approximately equal to cH_0 (the speed of light multiplied by the present value of the Hubble constant). Is this mere coincidence, or does it point to a deep connection between MOND and cosmology?

[2] Michael R Merrifield, 'Dark Matter on Galactic Scales (or the lack thereof)', in *Proceedings of the Fifth International Workshop on the Identification of Dark Matter*, edited by Neil J C Spooner and Vital Kudryavstev, pp. 49–58, World Scientific, 2005.

Chapter 7

[1] W de Boer in 'Indirect evidence for WIMP annihilation from Diffuse Galactic Gamma Rays', an invited paper at the 5th Heidelberg International Conference on Dark Matter in Astro and Particle Physics (DARK2004), Texas, October 2004.

Chapter 8

[1] Current thinking centres round the idea that an evolving quantum field, called the 'inflaton field', was responsible for driving the accelerating expansion.

[2] But see 'Perhaps dark matter is not so cold, Chapter 5, p.70, for an alternative view.

[3] Interestingly, the most recent 2dFGRS analysis, published in 2005, points to a somewhat lower overall matter density, $\Omega_M = 0.23 \pm 0.02$.

Chapter 9

[1] For a comprehensive review of the role of Type Ia supernovae in cosmology, see: Alexei V. Filippenko, 'Type Ia Supernovae and Cosmology' in *White Dwarfs: Cosmological and Galactic Probes*, edited by E M Sion, S Vennes and H L Shipman, Astrophysics and Space Science Library, Vol. 332, Springer (2005)

Chapter 10

[1] M S Turner, 'The Case for Lambda CDM', in *Critical Dialogues in Cosmology*, ed. N Turok, World Scientific (1997).

Chapter 12

[1] Volker Springel et al., *Nature*, vol. 435, pp 629–636 (2005).

[2] D N Spergel et al., 'Wilkinson Microwave Anisotropy Probe (WMAP) Three Year Results: Implications for Cosmology', submitted to *Astrophysical Journal*, 2006

[3] E Hubble, *The Realm of the Nebulae*, pp 201–202, Oxford University Press, London, 1936.

Index

Image Credits

Every effort has been made to acknowledge correctly and contact the source and/or copyright holder of each picture, and Canopus Publishing Limited apologises for any unintentional errors or omissions, which will be corrected in future editions of this book.

Unless otherwise stated below, all artworks are courtesy James Symonds.

Page vi: NASA, N. Benitez (JHU), T. Broadhurst (Racah Institute of Physics/The Hebrew University), H. Ford (JHU), M. Clampin (STScI), G. Hartig (STScI), G. Illingworth (UCO/Lick Observatory), the ACS Science Team and ESA
Page viii: NASA, ESA, S. Beckwith (STScI) and the HUDF Team
Page 2: SOHO (ESA & NASA)
Page 4: Nik Szymanek
Page 5: Faulkes Telescope Project & Nik Szymanek
Page 6: NASA, H.E. Bond and E. Nelan (STScI), M. Barstow and M. Burleigh

(University of Leicester, UK); and J.B. Holberg (University of Arizona)
Page 8 (left and right): European Southern Observatory (FORS/VLT)
Page 10 (top): The Royal Observatory, Edinburgh (original photograph taken with the UK Schmidt Telescope)
Page 10 (bottom): Robert Williams and the Hubble Deep Field Team (STScI) and NASA
Page 11 (bottom): Dr. Wendy L.Freedman, Observatories of the Carnegie Institution of Washington, and NASA
Page 12: The Royal Observatory, Edinburgh (original photograph taken with the UK Schmidt Telescope)
Page 21: Photograph from the Hale Observatories
Page 22 (top): this item is reproduced by permission of The Huntington Library, San Marino, California
Page 23 (top left): Jeffrey Newman (Univ. of California at Berkeley) and NASA
Page 23 (top right): NASA, ESA and J. Newman (University of California at Berkeley)

Page 29: Courtesy NASA/JPL-Caltech/A. Kashlinsky (GSFC)
Page 30: NASA, ESA, and the Hubble Heritage Team (STScI)
Page 33: Faulkes Telescope Project and Nik Szymanek
Page 35: Image courtesy of NRAO/AUI/NSF
Page 37 (bottom): Nik Szymanek
Page 38: The Royal Observatory, Edinburgh (original photograph taken with the UK Schmidt Telescope)
Page 41 (top left): NASA/CXC/E.O'Sullivan et al
Page 41 (top right): Palomar DSS
Page 41 (bottom): The Hubble Heritage Team (AURA/STScI/NASA)
Page 42 (bottom): Arecibo Observatory/ Cardiff University/Isaac Newton Telescope/ Westerbork Synthesis Radio Telescope
Page 44 (top): Courtesy of Prof. Duncan Forbes, Dr. Ale Terlevich and Dr. Richard Whitaker/the Isaac Newton Group of Telescopes, La Palma
Page 44 (bottom): European Southern Observatory

Page 46 (bottom): W.Couch (University of New South Wales), R. Ellis (Cambridge University), and NASA

Page 47 (top left and top right): S. Colombi (IAP), Canada-France-Hawaii Telescope Team

Page 48: ARC and the SDSS Collaboration, http://www.sdss.org

Page 50: European Southern Observatory

Page 51 (bottom left): NOAO, Cerro Tololo Inter-American Observatory; (bottom right): NASA and Dave Bennett (University of Notre Dame, Indiana)

Page 52: Nik Szymanek

Page 54 (left and right): The MACHO Project

Page 56: ESA. Illustration by Medialab

Page 61: © CERN

Page 64: Kamioka Observatory, ICRR (Institute for Cosmic Ray Research), the University of Tokyo

Page 65 (bottom left and bottom right): Tomasz Barszczak, University of California, Irvine, for the Super-Kamiokande Collaboration.

Page 67: Artwork courtesy of SNO

Page 68: 2dF Galaxy Redshift Survey

Page 72 (top and bottom): © CERN

Page 73: Ben Moore, Institute for Theoretical Physics, University of Zurich; www.nbody.net

Page 74: NASA, ESA, and the Hubble Heritage Team (STScI/AURA)

Page 75 (top): Romano Corradi (ING) and Laura Magrini (University of Firenze, Italy). Image courtesy of the Isaac Newton Group of Telescopes, La Palma

Page 75 (bottom): Ben Moore, Institute for Theoretical Physics, University of Zurich; www.nbody.net

Page 76: European Southern Observatory

Page 79: M Hilker (AIfA, University of Bonn); background: Michael Curtis Schmidt Telescope (A Karick, University of Melbourne)); insets: STIS/HST (PI: M. Drinkwater, University of Queensland)

Page 82: Francesco Arneodo LNGS-INFN

Page 83: Luca Cesaro/Alessandro Pascolini - INFN

Page 84: The Boulby Underground Laboratory and the DRIFT and ZEPLIN dark matter collaborations

Page 85 (all images): Fermilab

Page 90: The Boulby Underground Laboratory and the DRIFT and ZEPLIN dark matter collaborations

Page 91 (left and right): Nicolas Martin and Rodrigo Ibata, Observatoire de Strasbourg, 2003

Page 93 (top): ESA

Page 93 (bottom): ESA/J. Knödiseder (CESR) and SPI Team

Page 94 (top): ANTARES – F. Montanet CPPM/IN2P3/CNRS-Univ.Mediterranee

Page 94 (bottom): R. Muth

Page 96: Nik Szymanek

Page 97: NASA/CXC/F.K. Baganoff et al., and the H.E.S.S. Collaboration

Page 98: Ben Moore, Institute for Theoretical Physics, University of Zurich; www.nbody.net

Page 106: NASA/WMAP Science Team

Page 109 (top and bottom): The BOOMERANG Collaboration

Page 110 (bottom): CBI/Caltech/NSF (i.e. Cosmic Background Imager Project, California Institute of Technology, supported by the US National Science Foundation)

Page 111 (top left): CBI/Caltech/NSF

Page 111 (top right): NASA/WMAP Science Team

Page 111 (bottom): NASA/WMAP Science Team

Page 113 (top and bottom): NASA/WMAP Science Team

Page 114 (top): 2-degree Field Galaxy Redshift Survey

Page 114 (bottom): NASA/WMAP Science Team

Page 116 (bottom): NASA, ESA, J. Hester and A.Loll (Arizona State University)

Page 117 (top): NASA and The Hubble Heritage Team (STScI/Aura)

Page 117 (bottom): X-ray: NASA/CXC/ U.Illinois/R.Williams & Y.-H.Chu; Optical: NOAO/CTIO/U.Illinois/R.Williams & MCELS coll.

Page 119 (left and right): NASA, ESA, CXO and P. Ruiz-Lapuente (University of Barcelona)

Page 119 (bottom): NASA, ESA, The Hubble Key Project Team, and the High-Z Supernova Search Team.

Page 122: Reprinted with permission from "Supernovae, Dark Energy, and the Accelerating Universe", Saul Perlmutter, *Physics Today*, April 2003, pp 53–60, Copyright 2003, American Institute of Physics

Page 123: From R.A. Knop et al, *Astrophysical Journal*, 598, p102, 2003;

reproduced by permission of the AAS

Page 126: NASA, Adam Riess (Space Telescope Science Institute, Baltimore, MD)

Page 127 (top): NASA and J. Blakeslee (JHU)

Page 127 (bottom): NASA and A. Riess (STScI)

Page 134 (top left and right): WFPC2 image: NASA and J. Bahcall (IAS); ACS image: NASA, A. Martel (JHU), H. Ford (JHU), M. Clampin (STScI), G. Hartig (STScI), G. Illingworth (UCO/Lick Observatory), the ACS Science Team, and ESA

Page 135: Nik Szymanek

Page 136 (left): Kitt Peak National Observatory 0.9-meter telescope, National Optical Astronomy Observatories; courtesy M. Bolte (University of California, Santa Cruz).

Page 136 (right): Harvey Richer (University of British Columbia, Vancouver, Canada) and NASA

Page 155: NASA/WMAP Science Team

Page 156: NASA/WMAP Science Team

Page 157: Adam Myers et al. (University of Durham); background sky map: NASA/ WMAP Science Team)

Page 158 (all images): Dr. Volker Springel, Max-Planck-Institute for Astrophysics, Garching, Germany

Page 159 (all images): Dr. Volker Springel, Max-Planck-Institute for Astrophysics, Garching, Germany

Page 160: European Southern Observatory

Page 161: NASA/M.Markevitch et al.; STScI; Magellan/U.Arizona/D.Clowe et al.; ESO.

Page 162: European Southern Observatory

Page 163: NASA/CXC/IoA/S.Allen et al.

Page 165: Daniel Eisenstein, ARC and the SDSS Collaboration, http://www.sdss.org

Page 167 (top): VSA Collaboration

Page 167 (bottom): NASA/WMAP Science Team

Page 168 (top): CBI/Caltech/NSF

Page 168 (bottom): NASA/WMAP Science Team

Page 170 (top): Image: ESA [© 2002 ESA]

Page 170 (bottom): SNAP/Lawrence Berkeley National Laboratory

Page 171 (top): Courtesy LSST Corporation, http://www.lsst.org

Page 171 (bottom): Courtesy J.A. Tyson, U.C. Davis, and LSST Corporation, htp://www.lsst.org

Page 173: SNLS Collaboration

Page 174: SNLS Collaboration